IEE HISTORY OF TECHNOLOGY SERIES 27

Series Editors: Dr B. Bowers
Dr C. Hempstead

Restoring
Baird's Image

Other volumes in this series:

Restoring Baird's Image

Donald F McLean

Published by: The Institution of Electrical Engineers, London,
United Kingdom

© 2000: The Institution of Electrical Engineers

The Institution of Electrical Engineers,
Michael Faraday House,
Six Hills Way, Stevenage,
Herts. SG1 2AY, United Kingdom

British Library Cataloguing in Publication Data

A CIP catalogue record for this book
is available from the British Library

ISBN 0 85296 795 0

Typesetting and page layout by the author

Printed in England by T J International

Contents

Foreword

'Well what's the good of it when you've got it. What useful purpose will it serve?'

Frith Street, London on a cold wet January night in 1926, the group of eminent scientists from the Royal Institution had struggled up a series of dank narrow stairs to witness the demonstration of a new invention. On the whole, they were not impressed. Whilst most of them felt that the shrouded equipment hid some trickery, some went so far as to denounce the inventor as an 'absolute swindler'. Only a handful of the forty or so visitors that night had anything positive to say about what they had seen.

In his excitement and enthusiasm the inventor seems to have misread the reaction. Writing later he said, 'I was certainly gratified by the interest and enthusiasm. The audience were for the most part men of vision who realised that in these tiny flickering images they were witnessing the birth of a great industry.'

With hindsight we know that John Logie Baird was correct in his belief that it was the "birth of a great industry". We also know that no one (not even Baird) had any inkling of how great and powerful that industry was to become. With that same hindsight we can also begin to plot the changing public perceptions of Baird's achievements.

By the time BBC 405-line television started in 1936, Baird had travelled from obscurity to triumph to uncertainty. A few more months and he had begun the slide back into obscurity. For the next ten years he continued research, mainly into large-screen television and high definition colour where he produced some stunning developments. These, and many of his other television "firsts" such as stereoscopy, infrared, outside broadcasts, transatlantic television, video recording and colour go unrecognised by the public.

The mechanical television system he championed was, like its inventor, thrown on to the scrap heap by the mid 1930s, written off as an inferior, low-tech solution. It is rather surprising then that several elements of Baird's system resurfaced at CBS in 1950 (field-sequential colour), at Mullard in 1960 (the Banana tube) and on the moon in 1969 (Apollo 12 TV camera). Mechanical scanning is widely used today in fax machines, supermarket checkout readers and thermal imaging cameras where it simply provides the best engineering solution for the required task.

We understand the importance of these devices because we are familiar with them, with earlier inventions we have often to depend on museum collections to comprehend their status. Occasionally we come across an early technology that resists our attempts to understand it. Much of Baird's work has the problem that little supporting information exists to help us. He had a huge distrust of written records and what little was written down was probably lost in the Crystal Palace fire, which destroyed his laboratories and offices. For years the historic importance of these uninspiring looking Phonovision discs eluded us – until Don tracked down six of the discs and unlocked the key to reveal the latent images. Unexpectedly, the signals recorded on a handful of seventy-year-old discs took on a new significance and I became involved with Don and Eliot Levin in a project to transcribe them on to CD for archival preservation.

So it was that I arrived at Symposium Records in the summer of 1996 with the surviving discs. After we had discussed the technicalities of the transcription and the timetable for the next few days, Eliot dropped his bombshell - he had found another disc. The whole story behind the discovery and restoration of this, the Silvatone disc, is covered in the book. Suffice it to say that for the next couple of days we did not know what, if anything, was on the disc. The disc surface was so badly corroded that it looked unlikely that anything useful could possibly be retrieved, but Eliot transcribed it and Don took a copy away to work his magic on it.

Two days later I returned to collect the Phonovision discs. Don had made a first attempt at restoring the new disc, and with his laptop balanced precariously on a chair in the darkened corner of Eliot's living room I saw for the first time a real 30-line TV broadcast. All early television was ephemeral, live programmes, transmitted up the aerial and out into space. If you missed a programme there was no second chance. Here we had signals that by now should have been over three hundred million, million miles away in deep space but for some quirk of fate.

In the normal course of my work I have become used to the unexpected; out of several hundred new acquisitions each year there will always be a few surprises, but this discovery was different. Not only was the very existence of this type of recording a complete surprise, but also the fact that anything recognisable could be retrieved from such an unpromising disc.

The small flickering images I saw initially made little sense, (Don had not had time to do a lot of the subsequent restoration) but as we watched part of the sequence over and over, and as Don talked through his interpretation of the scenes, it all suddenly fell into place. With the hairs on the back of my neck literally standing on end, I realised how Howard Carter must have felt when the tomb of Tutankhamun was opened. Don and Eliot had just re-written a chapter in television history - the world's earliest domestic video recording had just shifted by over thirty years and provided us with an understanding of the style and quality of these pioneering programmes. For me, that first view of the Silvatone images was one of the main highlights of my professional career.

What John Baird would have thought of this one can only guess. He, more than most, would have understood the difficulties in replaying images from these discs. I like to think he would have approved of another Scottish engineer, living in exile, working alone to complete the work he started. Baird himself never received the true recognition he deserved, a situation that will doubtless be redressed over the years to come. Don has already started that process with his painstaking research into Phonovision and mechanical television. For me, his work is an outstanding example of industrial archaeology.

John Trenouth
Senior Curator of Television
National Museum of Photography, Film and Television
Bradford, June 2000

Preface

The restoration of early television images recorded on gramophone discs has increased our understanding of how television started in Britain. The discs have turned out to be time capsules of considerable historic value, made in the revolutionary age of early mechanically scanned television, dominated in Britain by the controversial, romantic visionary – John Logie Baird.

The earliest of the discs are shellac pressings made by Baird during a series of experiments to record a television signal onto gramophone discs. The aim of these Phonovision experiments was to develop pre-recorded videodiscs into a consumer product. The experiments were not successful and Phonovision was all but forgotten.

As a result of the recent restoration of a few surviving examples of Phonovision, we can now see in these fragments of video exactly what Baird saw, how he was trying to record television and what he was aiming to achieve. More importantly, we can now watch recordings made in 1927 and 1928, mere months after Baird had given the world's first demonstration of television, during his period of exploring this new medium and over a year before the first public broadcasts in Britain.

In the 1930s, soon after the first BBC Television Service had started, a few amateurs attempted to record the video signal onto aluminium gramophone discs. The discs lay unknown for decades, overlooked in the history of television. The recovery of the television images brings the first real evidence of just how good BBC Television was before the advent of high definition (405-line) television in 1936.

My discovery and restoration of the television images are set out in detail yet with the minimum of mathematics. A consequence of restoring the images is being able to isolate the faults. Studying these faults yields a wealth of new information on how Baird attempted the recordings and what problems he encountered.

However the restored images from Baird and from the BBC are in themselves a minor revelation. Most surprising of all is that the moving images of BBC 30-line broadcast television, not seen since they were transmitted in the 1930s, challenge long-accepted views that broadcasts were poor with amateurish performances.

This book seeks to explain how video recordings could be made 25 years before it was previously thought possible, why these recordings have been lost all this time, what type of television programme people sought to record, and what the BBC were doing with a Television Service in 1932, four years before television was supposed to have started.

Answers to these help understand the scope of these findings and help explain a fascinating yet poorly understood period in television history. An even greater understanding comes from drawing attention to the continued importance of mechanical imaging systems right from its practical beginnings in the 19th century to the present day. By taking a vantage point within a new age of digital communications and the Internet and at the end of the analogue television era, the book presents a fresh look at the development of television. This allows the reader to appreciate the context in which the historic video recordings were made and what led to their creation.

Though it involves computers and television, this story tells of an unwitting foray into a form of archaeology, delving into a decade-long *dig* into Britain's engineering heritage. Unlike traditional archaeology, the artefacts are not embedded in layers of history but have existed in both private and public collections, largely ignored as curiosities. The tools to uncover and restore are not trowels and brushes but specially developed computer-based signal and image processing. The period is not some dark age in Britain's past; it is the early part of the twentieth century, a time between two of the worst wars in history and a time of great technological change. The task of restoration is not an occupation or profession, it is the result of my fascination and enthusiasm, fuelled by results.

The story begins in 1981 with my recollections of how this unusual dig started. I had borrowed a documentary on audio LP disc from Harrow library. It was a light-hearted history of television narrated by the comedian, John Bird, and called 'We seem to have lost the Picture'.[1] Though mildly entertaining and, as it turned out, wildly inaccurate, it included something utterly fascinating. At one point, Bird introduces a strange sound, describing it as 'Baird's brain-damaging buzz-saw'. Sounding more like a swarm of angry bees, this was supposedly a recording of the vision signal from Baird's original 30-line television system. Finding this fragment was timely: I had just completed building the software to capture audio into my home-built computer and here was something to work on.

On a sunny Saturday morning in late 1981, I was huddled on the floor of my spare bedroom in Northolt with tape deck, hand-built computer and oscilloscope, a tangle of wires and a tuneless chorus of various cooling fans. The green flicker of the oscilloscope trace was difficult to decipher. I was looking at what was supposed to be a video signal. I could see that the waveform repeated in a slowly changing pattern every 80 milliseconds, and another pattern repeated within it. This was undoubtedly a signal from out of history: a 30-line television signal with a picture rate of 12½ per second.

Though documentaries and books had covered this early version of television in detail, there had never been any mention of early video recordings. Surely those recordings were worthy of at least some comment? I wrote some software to unravel the data, stored it in the computer's memory and plotted it as a 30-line picture. Against expectations, I saw the head and shoulders of a cartoon-like drawing of a man that looked vaguely like Charlie Chaplin. Though the picture was distorted and unstable, it was quite a surprise to see anything at all from the disc. How amazing! A picture embedded in the sound! The character did not move, but then the two sequences on the LP were each only a few seconds long. I cleaned up the picture, sat back and asked the question, 'I wonder if there's any more?'

---

[1] 'We seem to have lost the Picture' (BBC LP from the series '40 Years of Television'), 1976, REB239

Acknowledgments

The discovery nature of this book means that a great deal of the material presented is original and new. In addition, a considerable amount of supporting material puts the story into context with the history of imaging technology. Where possible, this material has comprised facts and photographs that are either new or have rarely before been published.

I am particularly indebted to Clare Colvin of the Royal Television Society (RTS), John Trenouth of the National Museum of Photography, Film and Television (NMPFT), Nicholas Moss of the BBC and Ray Herbert for supplying photographs to use in this book. In addition I thank all those referenced for permission to use their material freely. In maintaining the book's theme of restoration and preservation, I have suppressed the defects of age, such as shading, scratches and crease-marks, in every one of the these historic photographs using off-the-shelf computer graphics software.

Over the years, various people have supplied me not only with historic information but also that valuable commodity for a solo worker, encouragement. Foremost among them is Ray Herbert, whom I have known since 1982. Ray is a former Baird Company employee whose personal knowledge of Baird's achievements and whose personal archives of the Baird Company are unsurpassed.

Time has taken its toll in the years since this work began; almost all of the pioneers of British television engineering from the 1920s and 1930s are now no longer with us. Amongst those, Tony Bridgewater showed probably the greatest interest in this work – particularly since he had been a Baird employee who transitioned over to the BBC for the 30-line Television Service. I remember Bridgewater and the other pioneers with respect and gratitude for their encouragement.

My thanks also go to Doug Pitt and all the members of the Narrow Bandwidth Television Association (NBTVA). They keep alive the interest in those early days of television through their innovative use of technology to achieve television on amateur radio channels.

This book would not have been possible without the kind support from the owners of the discs. I am grateful to H. C. Spencer, J. G. S. Ive, B.

Clapp and E. G. O. Anderson for allowing me in the 1980s to transcribe their precious discs. I hope I have made them even more precious by doing so. At time of writing, three of the six Phonovision discs are in the collection of the NMPFT in Bradford, one is in the archives of the RTS and the two remaining discs are in private hands.

For the later recordings of BBC television transmissions, I am grateful for Dave Mason and Jon Weller for bringing their discs to light. I am especially grateful to Eliot Levin of Symposium Records who provided the essential transcription of those later discs and undertook the archival transfers of all the Phonovision discs for the NMPFT in 1996.

The historical background on John Logie Baird was drawn in part from his autobiography. I thank Professor Malcolm Baird for permission to quote extensively from his father's book. In addition, I thank Earl Sherrin for permission to quote from J. D. Percy's unpublished memoirs.

In preparing this book for publication, Martin Salter of R. H. Consultants (lately of Ampex Corporation) and Ray Herbert helped considerably by applying their extensive knowledge to check and correct the book for historical accuracy.

My thanks go to Martin Eccles, Editor of 'Electronics World', for permission to re-use the title of my October 1998 article, 'Restoring Baird's Image' and for supporting my series of five articles.

In gathering the photographic and written material that supports this book, efforts were made to track down the original copyright holders. In a few cases it may be possible that copyright has been incorrectly attributed. In that instance, I apologise for the error.

I dedicate this book to my children Rhian, Malcolm and Heather, and to my wife, Lydia – the first to share in these images from the dawn of television.

1   As Others See Us

*'Oh, wad some Power the giftie gi'e us, to
see oursels as others see us.'*

Robert Burns, 'To a Louse'

Videodisc comes of Age

It had been a busy day in the City, but he had grabbed some time to drop by
the West End store that was selling the new discs. The videodisc he wanted
had just been released that day and he was going to be first in his
neighbourhood to get it. This was his favourite singer after all, and he
wanted to see how she came across compared with the performance she
gave in the West End a while back. He pulled the disc carefully out of its
sleeve and placed it in the player. Still with the fascination of this latest
wonder of technology, he watched as the disc ran up to speed (see Figure 1-1). The music rose gently and soothingly from the loudspeaker and the picture burst into life. There she was! Betty Bolton was just as he remembered her in performance. That straight black hair shining in the studio lights, the little curl that was almost her trademark and even the tiny silver clasp that held her hair just so. The low-slung dress and the necklace of dazzling stones that twinkled as she turned. He took it all in.

This was a slow love song and her expression of delight and her actions came across so well, showing that she was really living the part. The picture itself was clear, and absolutely rock steady. Surely, he thought, this was better than the

Fig 1-1. A promotional mock-up of the 'Phonovisor'. The image would appear in the aperture to the left of the turntable.
Courtesy of the Author and Betty Bolton

BBC Television programme he had seen her in. Now of course he could play this music video anytime he wanted. He didn't have to wait until late at night for the television programme. He hated being quite so tired and bleary-eyed in the office the next morning.

As he watched, the neon light gave the image a curious realism and depth as the image in the big lens showed the attractive performer at her most alluring. A quick glance at the playback head showed him he was most of the way through the recording. The special playback head was unlike his record player. This could play back both video and sound from the disc at once – and just from one needle in a groove. The disc finished and he removed it, carefully returning it to its sleeve just as his wife entered the room.

'Oh no, you haven't been wasting more money on that new-fangled contraption, have you?' she said in a timeless manner.

'But dear, it's the latest thing and it didn't cost much more than the normal sound disc'. He held back from saying that the Baird videodisc player was about the same price as a top-of-the-range electric gramophone player – and it came with its own integral TV display. Excellent value, he thought. 'After all, my love', he added. 'This *is* 1934 and we must keep up with the times'.

This little story could have so easily been fact. There *could* have been a videodisc system available in the 1930s. It would have been a novelty, able only to play short four-minute programmes, and an adjunct to the emerging high-volume market for shellac audio discs. From its original concept by John Logie Baird in his patents of the late 1920s, this would have been the simplest and cheapest combined videodisc player and display system ever built. Not only that but this domestic novelty of pre-recorded videodiscs would not have been bettered until the 1970s, some 40 years later. For reasons that only recently have become clear, Baird's machine never appeared and there is one less failed commercial videodisc format for history to record.

Even before 1934, television was a reality, though limited in scale and budget, and constrained to the technology of the day. The broadcast services were similarly limited, being implemented with some elements that were mechanical rather than electronic. To some of us, the thought of mechanical television seems bizarre, yet mechanical scanning is at the heart of many of today's systems of image formation and transmission.

We use it every day when we feed a document into a fax machine. Our weather is forecast using satellites that provide mechanically scanned images. The latest military reconnaissance images in thermal infrared are scanned mechanically just like those early television images from the 1930s.

Mechanical Television Today

All weather satellites – the geostationary and the polar orbiting – use mechanical scanning to build up their pictures. It was not always like this. The first weather satellites used electronic television camera tubes to take cloud-cover pictures. With the advent of the Tiros-N series of satellites in the late 1970s, those satellites used mechanical scanning as a *more advanced* arrangement. This seems completely wrong. Surely mechanical scanning is supposed to be yesterday's technology? There were two excellent reasons for progressing to mechanical scanning. First, mechanical scanning can generate extremely high-resolution images – far higher than a television camera can achieve. Second, weather satellites need to be able to *see* not just in visible light, but also in the far-infrared, typically at 4 and 10 microns wavelength. Fabrication is difficult and expensive for sensors at these wavelengths. As of today, it is still more cost-effective to build a single photo-sensor and mechanically sweep the image across it.

It is in the area of military reconnaissance and surveillance that we find the latest and best imaging technology. Thermal infrared television is central to battlefield imaging systems. Seeing the enemy day or night from the heat generated by their vehicles and their own bodies, gives a tremendous tactical advantage. Traditional camouflage becomes ineffective.

In the 1970s, GEC of Basildon, Essex, developed a thermal infrared television camera based on a short stack of photo-sensors. The camera was called TICM (Thermal Imaging Common Module), pronounced 'tickem'. Since then, these thermal imagers have been adapted for almost every military use: in the battlefield, on board ships and even fitted on aircraft for capturing reconnaissance imagery. The unusual feature of its construction is the use of mechanical scanning based on the mirror drum – the same technology solution used by John Logie Baird.[1]

Mechanical Computers?

Though we live in an electronic age, our computers run only because of the electric motors that spin discs (whether floppy-disc, hard-disc, CD (Compact Disc) or DVD (Digital Versatile Disc)) and the mechanical sweep of the read/write heads across the disc surface. Those discs store our precious data and applications necessary to run our computers. Though semiconductor memory is becoming cheaper and denser, rotational disc technology still provides the most cost-effective solution to storage ... for now.

For nearly 50 years, the primary method of recording video has been by using tape rather than discs. Tape is for high capacity data storage, whilst disc is for rapid access to the data. The method of magnetic tape recording today is much the same as it was nearly 50 years ago. For every machine that uses videotape – from the original broadcast machines to the digital video recorders and camcorders of today – magnetic tape is wrapped around a drum and the heads in the drum are spun at high speed across the tape surface. Mechanical scanning in some form or another is still central to our daily existence.

'Mechanical' Space Imaging

Even throughout the space age, mechanical scanning was in many instances the method of choice. From the second moon landing, Apollo 12 onwards, the activities of the astronauts on the lunar surface were watched on a colour television camera. The colour image came from a motor-driven colour filter wheel, spun in front of the electron tube camera. Further out, the first probes to distant planets chiefly used mechanical scanning, as it gave the simplest and the highest quality image. The Pioneer probes to Jupiter and Saturn (pioneering the way for the Voyager spacecraft) used mechanical scanning to build up our first close-up images of these planets and their moons. The pictures from the first landings on Mars in 1976, Vikings 1 and 2, used pure mechanical scanning via nodding mirrors to build up high quality images. By the time of the next successful Mars landing 21 years later, technology had moved on. The now-familiar vista of rocks and sand was captured with a solid-state digital chip camera. The electron tube has never had a *look-in* on the surface of Mars.

Seeing at a Distance

On 4th July 1997, a United States spacecraft made a type of landing on the surface of Mars that no human will ever make. In the dark hours before Martian sunrise, the spacecraft plummeted through the thin atmosphere of the red planet and hit the rocky surface at up to 90 km/hr (56 mph). The spacecraft hit and rebounded 12 metres off the surface, bouncing a further 14 times. It came to a rest, deflated its cocoon of protective balloons and opened itself up like the petals of a flower. The Jet Propulsion Laboratory (JPL) and the National Aeronautics and Space Administration (NASA) received colour images from the two digital cameras on the lander – Mars Pathfinder – and immediately made them available on the World-Wide Web (see Figure 1-2). The interest in seeing pictures in near real-time from the surface of Mars was overwhelming. NASA called in private industry to help feed the demand for pictures from around the world. When the little robot rover – Sojourner – rolled down its ramp onto the Martian surface on 8th July, the distributed network of 50 computer servers clocked up an impressive 47 million *hits* in just one day. At the time, it was the biggest ever Internet event. By 4th August, the total number of hits had reached 566 million.

Why was there such an unprecedented interest on the Internet for Mars Pathfinder? The spacecraft was a low-budget unmanned scientific explorer. In many ways it was inferior to the pair of Viking probes that had soft-landed on Mars exactly 21 years before. The JPL and NASA had provided the closest approximation to having television from Mars. The popularity of the pictures on the web site was due to its immediacy. As we studied the pictures appearing almost *live*, we were *there* on the surface of an alien world seeing what the camera was seeing just a few minutes before. This is the fascination of 'seeing at a distance' as an event unfolds.

The World-Wide Camera

The Internet is also home to web cameras that show, live, the view from hundreds of digital cameras around the world. Though Orwell predicted something like this in '1984', the situation today is less 'Big Brother is

Fig 1-2. An enhanced photo-mosaic from one of the digital chip cameras on board the Pathfinder spacecraft on the surface of Mars.

Courtesy of NASA/JPL

watching you' and more like members of the Global Family watching their relatives. Fortunately for us, Big Brother *is* watching, maintaining public safety in car parks and malls using the television eyes of surveillance cameras. Those eyes continually capture the scene on videotape for the occasion when it might be used as evidence in court.

We can see in the space of a few minutes of surfing the web, afternoon traffic in Arizona and Chile, sunset at DisneyWorld Florida, early evening in Buenos Aires, Argentina, the face of Big Ben at night, the Kremlin and Red Square in the small hours of their morning, sunrise in Singapore, rush hour traffic in Sydney and lunch-time traffic in Honolulu.

If we really felt like 'seeing at a distance' – the literal meaning of the word 'television' – the Internet also holds recent all-earth views from 22,300 miles out from five weather satellites spread around the globe in a Clarke (geo-stationary) orbit. The outstanding quality of the raw pictures, across 12 different spectral bands, each comprising as much detail as in 140 television screens (64 million values), comes from the use of mechanical scanning to build up the picture (see Figure 1-3).

The Internet in 2000 gives us coverage of over 3,000 television and radio stations around the world. Many of the television stations are live relays reduced in detail and quality to allow moving pictures to be received on a domestic telephone line. Those live images have the fascination of allowing us to see the output from not only the main European and US networks but, for example, 'Chilevision' from Santiago in Chile or Sarimanok News Network in the Philippines. The heavy compression generates crude pictures that resemble the first television images of the 1920s. We gain some idea of the excitement television caused when people saw it for the first time, from the novelty in being able to graze around the crude offerings from world television via the Internet.

This revolution in 'seeing at a distance' comes not from any major development in television, but from the incredible growth in global communications and computer connectivity via the Internet. Since the mainframes of the 1960s, the minicomputers of the 1970s, the first home computers of the 1980s, we have seen computers become an integral part of our society and culture. They form the backbone of the world's financial, commercial, government and military operations. Supported by high volume, low cost communications, the computer has led the Internet revolution, with the potential for connecting everyone on the planet with a computer and telephone connection.

History in 'Sound-Bites'

The technology and, more impressively the broadcast content of radio and

Fig 1-3. The view from Clarke (geo-stationary) orbit. These mechanically scanned images were all taken at mid-day GMT on the Spring equinox, 2000. With new images sent down to Earth every 15 minutes and the slow pace of sun and cloud movement, this is 'live' seeing at a distance – essentially 'television'. The five satellites from top left to bottom right are GOES West, GOES East, Meteosat (split across two rows), Indoex and GMS.

From originals courtesy of Dundee University

television capture the imagination and stir nostalgic feelings. This deep interest is somehow enhanced by the entire history of radio and television broadcasting lying almost within a single lifetime. Events that occurred in those early days ought to be recalled with some degree of accuracy. That is true for questions like, 'Where were you when Kennedy was assassinated /

man first stepped on the moon / Princess Diana was killed?' But these are special world-class events that cause us to take a snapshot of our lives rather than remembering the events themselves.

Although early television elicits clear memories in people of what they were doing, watching and listening to, knowledge of how television developed is usually sketchy. It is a complex subject, involving rapid changes in technology, a lot of which the public were not fully aware.

Our ignorance of the subtleties is understandable and excusable: we have, after all, better things to do. Indeed the broadcast companies who re-tell the story of television assume a short attention span and fit the story accordingly. The problem is then that this 'sound-bite' version of history *becomes* history, 'dumbed-down' to such a state that there are glaring anomalies and misunderstandings.

Here is an example. 'John Logie Baird invented Television. Television started in 1936. There was a competition for the best system and Baird's mechanical 30-line system easily lost out to EMI's 405-line system.' Not one of those statements is strictly speaking correct. Far more serious is that those statements came from a TV producer preparing for a documentary on early television. Some of these errors are understandable because of the complexity of television's history. Apart from these errors, there is also a perception that the old part-mechanical technology for television is wrong.

The discoveries made from the restoration of videodiscs, made decades before the term was coined, are a minor revelation. They show quite clearly that there has been a migration of thinking. Over the years, the achievements in the first decade of television have gone from being headline news, to being praiseworthy, to attract derision and then to being ignored. In those early days, John Logie Baird brought Britain and the British public into the television age. The story of this early dawn of television is worth telling and worth getting right.

When it comes to such an area of relatively high technology, the challenge is to be able to gain an understanding of not just what the technology is and how the solution came about, but what it means and what it offers to us. All too easily, the sheer confusion of technology means that we get immersed in the details of how the technology works. Not only does this make us more remote from the solution but it also creates its own culture and develops its own language.

Struggling with the Language of 'Techno-speak'

Technology is moving faster than the language used to describe it. We have grown to accept that technology developments are a way of life, and expect improvements, yet we are becoming increasingly distant from it. We are not exactly short of words in the English language, yet we appear to have run out of new words – for the technology at least. Our struggle with the language for technology is a good example of our poor understanding of what it does for us and to us.

Technology to bring the media to the people, such as hi-fi, radio and television affects most of us. New developments and products attract either trade names (such as Sony's Walkman, Philips' Laserdisc), acronyms (such as VHS, NICAM, PAL, DVD) and abbreviations (such as camcorder, TV, video, the 'tube').

In the 1960s, we used to talk about *transistors* not as the electronic component but as an abbreviation of portable transistor radio receiver. We all listened to transistors in those days. That did not mean we all stuck three-legged electronic components in our ears. The term 'transistor' has dropped from popularity and been replaced simply by 'radio'.

Nowadays, television has its fair share of word over-use. The most notorious is *video*. The word is Latin, meaning 'I see'. Video – as in vision signal – crept into the engineers' vocabulary after the Second World War. Video has entered into common use with such vigour that it is applied as noun and verb. 'I'll video that for you' refers to recording on videotape. The noun 'video' is either a recorded videotape or it is the video recorder. So we might conceivably get someone saying, 'Put the video in the video and I'll video that programme for you.'

It is the use of this special language that is one of the causes of misunderstandings and myths. For instance, the often-used statement 'Television started in 1936' should *really* be 'High-Definition (not less than 100 lines per picture) television broadcasting started in Britain in 1936 with coverage of the Greater London area'.[2] It is a bit of a mouthful, but at least a more accurate mouthful.

What does 'Television' *really* mean?

'Television' is literally 'seeing at a distance'. The word has been around for some time. It was first used in 1900 in Paris, France. At the International Electricity Congress, Constantin Perskyi, later Professor of the St Petersburg Academy of Artillery, presented a paper entitled 'Television'.[3] Though the content was hardly memorable – relating to a mechanically scanned system – the title of the paper caught on rapidly. To put this in perspective, this first use of the word was at least twenty years after the first solutions to television had been patented and fully a quarter of a century before its first demonstration by John Logie Baird.

The word 'television' is now used right across every aspect of the medium: from the industry itself, to the occupation, the technical system to create it, the art form, the programmes we watch, right down to the 'TV' – the 'box' on which the pictures and sound appear.

When we ask engineers what television means, they'll talk about the number of lines, the display rate, the colour system (NTSC or PAL). If they are talking about the old systems, they may even indulge us with terms such as back porch, sync pulse, front porch, colour burst, mixed blanking and field sync. If we ask a digital television engineer, we will find similar terms, but interspersed with the language of the computer and digital communications technology: data packets, correlators and memory buffers. When we ask the programme makers, they will talk about the message, the material, the mission and, most importantly for them, the ratings – how popular their creation was.

But none of this – the technical, the programmatic, the social message – is strictly the original meaning of television. These are all solutions to it, implementations for it or implications from it. All these views miss the point – television is 'seeing at a distance'. Like radio, television is a medium for communicating and also like radio, there are many different and valid ways of doing it.

This mistaken identity for television has been the root cause of confusion in trying to write its history. To the question 'Who invented Television?' we should be responding, 'What do you mean by "Television"?'

---

[1] MAINES, J. D.: 'Surveillance and Night Vision', *IEE Electronics and Power*, Sep 1984, pp679–683

[2] BSI 205-1936, subsection 108–109

[3] LANGE, A.: Private Communication with Author, 28th Feb 2000

2   Distant Vision

'If I have seen further, it is by standing on the shoulders of giants.'

Attributed to Sir Isaac Newton, 1675

Revolutionary Television

In the grand picture of the achievements of humankind in the 20th century, there are innumerable examples of systems that have changed the way that we live. We have seen those systems grow from early beginnings to maturity within the space of a single human lifetime. In terms of both development of the technology and the sociological changes it brought about, television ranks as one of the most significant.

On 23rd February 2000, the United States National Academy of Engineering announced its top 20 Greatest Engineering Achievements of the 20th century. They placed Radio & Television in 6th position, behind Electrification, the Car, the Aeroplane, Water Supply & Distribution and Electronics. At the press conference, Professor Neil Armstrong said, 'Engineering helped create a world in which no injustice could be hidden.' Interestingly, the engineering that landed him on the moon and explored our solar system and universe came in 12th place.

Supported by the products of the Communications Age, television's ability to bridge continents and bring into our homes images that can educate us, inform us, make us laugh or cry, horrify us, or even fool us, is unique. The power of the picture is awesome and we have seen how the use of television can be manipulated for major political influence, and how its graphic message can even play a major part in stopping wars, notably Vietnam. Indeed President Lyndon Johnson said on that last point, 'historians must only guess at the effect that television would have had during earlier conflicts'.[1]

From Scratch...

Though the desire for television is simple and ages old, the development of it is complex, immersed in detail and above all recent, rapid and still in progress. What we have today is only an *implementation* of television defined and constrained by the available technology.

There are widely different views on television with some even arguing that it is already dead[2] This comes from taking a culturally centred view and ignoring the real performance differences between the demand-based Internet and broadcast television. The image quality of video across the Internet is currently poorer than that of Baird's 30-line system despite the technology being used to achieve it. This is no more than a consequence of not only the low performance of the computer connection to the Internet but also congestion of the network caused by having as many video feeds as there are demands for them.

In our future, television may remain a broadcast service, become a demand-based service or, most likely, end up as part of an integrated home entertainment system. None of these detracts from its short and feature-packed history. The future is for speculation, but the past can be both fascinating and informative – especially when trying to understand how we got to the present, and how we can fulfil the dreams of the thought pioneers of the 19th century. Though what we have today is beyond their wildest imaginings, those pioneers also dreamt of possibilities that have yet to be realised. We tend not to think of some of those early ideas as being anything to do with television. In fact, the dreams and ideas that allow us to *see at a distance* came very much earlier and had a much greater extent. Those dreams incorporated all methods of imaging whether for capturing and recording the image of a scene, or whether for sending that image over long distances. Television is merely one aspect, one implementation of imaging.

When we try to understand how television started – especially in the context of the *gramophone videodiscs* – we need to deliberately avoid the temptation to roll up our sleeves and start talking about specific television developments. We must stand back from it, and start thinking 'outside the box' to get television placed in proper context with other imaging developments. From that position we will be able to see what drove those developments, where the ideas came from and what the expectations were.

How it's Done

The desire for seeing at distance is probably as old as humankind itself. To be able to view some representation of a scene in a remote place has captured imaginations through the ages. No one person can take credit for the end result since the development of the concepts for *distant vision* took place progressively. The picture that developed over the ages is now the hallmark for all vision systems. It comprises four elementary stages: a *camera* stage converting scene into image, a *transmission* stage that sends the image, a *reception* stage that receives the image and a *display* stage for converting that image back into a reproduction of the scene. These four stages form a generic definition for distant vision. For completeness, there is an additional stage to allow for capturing or recording the scene information.

This definition is so basic that it can apply to every known type of situation involving images, from photography to facsimile, to the movies and television in all its forms. The difference is in the detail.

The table below shows the mechanism at each stage for each of the different types of imaging.

Imaging	Camera Recording Equipment	Sending and Receiving	Recording System	Display System
Eye	Retina: rods and cones	Optic nerve	Memory	None: fed into perceptual process
Photography	Glass plate, Celluloid film	Distribution of reproductions	Inherent in format	Hard-copy, Projector
Movies	Celluloid film	Distribution of reproductions	Inherent in format	Film projector
Facsimile	Mechanical scanner	Phone line transmission	None	Mechanical printer
Television (mechanical)	Opto-mechanical scanner	Radio (medium & short wave)	Experimental audio cylinder, audio disc	Opto-mechanical scanner
Television (electronic – analogue or digital)	Valve-based camera tube scanned electronically, Solid-state imager (CCD)	Radio (VHF, UHF) Direct Satellite	Videotape, Videodisc, Computer disc	Cathode-ray tube (valve-) based display LCD for portables Plasma screen

Table 2-1. Comparison of the various methods of imaging.

From Scene to Image

The first stage in all of these different systems is the camera, and the first stage of the camera is common to all types of imaging, is the simplest, the best known and you're using it right now. The lens in the front of the eye forms an image on the back – the retina. Our complex surroundings are reduced by the eye's lens to a flat (two-dimensional) scene of fixed coverage. The eye with its retina, optic nerve, visual cortex and the rest of the brain form our own personal vision system.

That first step of forming an image is probably the simplest and the easiest to ignore. Getting it right is absolutely crucial to all distant vision systems.

Easily the most impressive vision system I have ever seen was as a very young boy on a day-trip to Edinburgh. Still vivid is the memory of the clarity and sheer detail in that big round picture display. There was no flicker, the movement was smooth, and the scene was filled with vibrant natural colours topped with crisp photographic detail. This was Edinburgh's 'Camera Obscura' (see Figure 2-1). Situated near the top of the Royal Mile on Castlehill, just down from the Castle, the Camera has a good clear view over most of the Capital. It is at the top of Outlook Tower, up about six flights of stairs. The Camera is a leftover Victorian novelty that probably had as much impact on Victorian adults as it did on me as a child. It first opened to the public in 1853 and had its optics refurbished once in 1947. It is still a popular attraction due to its key position just down from the Castle.

Fig 2-1. The 'Camera Obscura' in Edinburgh viewed from the forecourt of the Castle. The 'Camera' view takes in most of Edinburgh and the surrounding land. The steerable aperture is mounted on the top of the dome structure.

Courtesy of the Author

The Camera is no more than a lens system that forms the optical image – the first stage of distant vision. The Camera Obscura is an elaborate form of the pinhole camera. Most of us at one time will have built one: a sealed biscuit tin (minus the biscuits!), a pinhole in one side and greaseproof paper stretched over the other side. Shading this other side, we would see a faint upside-down image on the greaseproof paper.

The principal is so simple and basic that it is difficult to attribute an inventor to it. The Camera first appeared in print in a book by Frisius in 1545 but may even have been demonstrated and used over two millennia before by Aristotle.

In the late 17th century, a Jesuit named Kircher reversed the principal of the Camera creating a projector. This became the 'magic lantern' of the 19th century, eventually evolving to today's slide and movie projectors.

By the early 1800s, the Camera Obscura appeared in portable versions, primarily for artists on the move. Those portable cameras provided the design basis for the first photographic plate cameras.

Photography

The next stage in distant vision converted the flat image to a picture. For photography, the key development was finding that certain mineral salts darkened when exposed to sunlight. In the late 18th and early 19th centuries, chemists worked to produce a flat photographic glass plate coated with those salts. The challenge for the chemists was to create a process that would first rapidly form the image during exposure, but then change its properties to allow viewing of the image in strong light. They were wildly successful and today, after 150 years, we can still view these images almost exactly as they first appeared.

The photographic plate or film is a capture device, storage device and display device – all rolled into one (see Figure 2-2). George Eastman introduced both the name 'Kodak', a box camera already loaded with film, and a flexible transparent film paving the way for mass production of stills, and more importantly, the *movies*.

For photography, the technology challenge was in the chemistry of the materials and processes to create the image. That technology has been mature for decades. One challenge remains for the chemists today: finding a truly permanent colour process. As cheaper processes were introduced around the 1970s, colour took over from black-and-white photography as the most popular medium for photography. However, the dye-based process used for colour fades in time. Today, if we want an image to last the test of time, it must be shot in black-and-white.

The developments in photography have centred on the camera, the materials for its construction and the electronics to help take a properly exposed picture. The basic principle of the Camera Obscura and photo-sensitive film has remained unchanged since the mid 19th century. Of all the imaging methods, photography can still be just as effective without any electronics. For some photographers, the best camera is one that is manual and mechanical.

Fig 2-2. The 'Vest Pocket' Autographic Kodak.

Courtesy of the Author

Now, we could say that Photography has little to do with Television. Recently however, those technologies have converged with the appearance of cross-over consumer products. We need only look at digital cameras to realise that there is very little difference between the cameras (primarily for Photography) and Video Camcorders (primarily for Video Recording). The late 1990s saw the advent of Camcorders that could also record digital stills on its videotape and Digital Cameras that could also record short video sequences with sound onto its storage chips. We find it so easy to keep Photography and Video Recording separate in our minds – but they are simply different technology solutions to capturing images.

Facsimile

Facsimile is the general term for reproducing copies of photographs and documents at a distance. The fax machine in our office or home is just one solution to this. Yet, its beginnings lie in the Victorian era well before electronics and light detectors and long before television.

Sending images without using light sounds a bit of a challenge. Instead of measuring light reflected back off a document or picture, those earliest of facsimile systems used electrical conductivity.

Having the material prepared for transmission could be a major task in itself. The drawing, photograph or text was etched, indented or otherwise embossed on a smooth metal plate. As the rod brushed over the surface, connection was interrupted by the plate's pattern. The resulting current would be either 'on' or 'off' depending on the pattern at the point of contact. The electrical circuit was no more complex than a battery with one connection attached to the patterned metal plate and the other to the metal rod.

The rod had to be swept evenly – scanned – across the entire surface to fill out the picture. By having the same scanning system at the receiver, by using paper instead of the metal plate, by having a pen instead of the rod and a means of lifting off the pen depending on the current, the system could create a paper copy of what was being transmitted.

The surprise comes when we realise just when such a system was first thought out. It was a Scottish clockmaker, Alexander Bain, who, in 1843, was the first to propose a use for the telegraph using a system similar to that above. His patent dealt in theory with most of the aspects of distant vision via facsimile: scanning to paper wrapped around a rotating drum and automatic synchronisation – ensuring that transmitter and receiver were perfectly in step.[3] Just a few years afterwards, such a system became practical. It was Bain who suggested scanning an image line by line, who indicated that synchronisation of transmitter and receiver was essential and who suggested transmission of an image could be achieved along a single wire. Bain's invention is the taproot of all electrical imaging systems, assuring his recognition as the father of image scanning and facsimile reproduction.

The first such systems created a binary-level picture. The two brightness levels occurred where there was either ink or no-ink – a digital image of sorts. Our conditioning on digital technology makes us think of digitally transmitted images as recent. Rather, they came *first*.

Of course, this is not very surprising. Before it became practical to have the telephone, we used the telegraph – an inherent digital communications system. The telegraph was based on switching electric current flow through cables. Various coding systems were used to convert and send messages – the most famous and enduring of these being the Morse Code.

Shedding Light on Selenium

In 1873, the proverbial door was opened for the first step to be taken in real-world imaging. Through pure accident, something was found to be very strange about the element Selenium.

Selenium was a relatively new find: it was discovered in 1817 by J. J. Berzelius, Professor of Chemistry in Stockholm, Sweden. Although it was a rare element, rare enough to avoid detection until then, it appears almost everywhere on the planet. More recently, we've discovered that Selenium is one of those trace elements the body simply cannot do without – even though the average amount in a human body is around 14 thousandths of a gram.

In 1873, Willoughby Smith was working at the Telegraph Construction and Maintenance Company. He had already made his name in developing insulated electrical cables for the latest technology of the day – telegraph communications. He had managed the process of making and laying the submarine telegraph cable between Dover and Calais in 1849 and had played a part in the first transatlantic cables laid in 1865–66.

Working for him at that time was a young engineer called Joseph May. In 1873, Smith and May had been experimenting with using Selenium as an insulator for a new idea that Smith had about electrical insulators for submarine cable.

Smith was puzzled by a strange variation in his measurements of the degree of insulation. However, May spotted that when light fell on the Selenium, it stopped behaving like an insulator and started conducting electricity. Not only that, it conducted more electricity the more light fell on it. By accident, Smith and May had discovered a method to convert changes in light to changes in electricity.

The discovery acted as a catalyst for new ideas for 'electric images'. It led inevitably to major new developments in facsimile transmission and in the evolution of thinking in how to achieve television.

Selenium itself was a poor light detector for imaging. It did not react instantly to changes in brightness. Yet, it was the only source available for some time. The slowness of its response meant that scanning the pictures took tens of minutes – even hours. Even at these speeds, electrical transmission was a lot quicker than sending the pictures by mail.

News Facsimile

Facsimile was adopted early on as the primary means of transmitting news photographs. The techniques were far advanced by the start of the 20th century (see Figure 2-3). The main challenge affecting all methods of sending images – synchronising the image scanner with the reproducer – had been overcome.

Facsimile received a further step improvement with the introduction of valve electronics. This allowed great improvements to be made in quality and performance. Very long-range trans-missions of continuous tone (i.e. smooth grey-scale or analogue) photographs could be made. Major events could be communicated around the world just a few hours after they happened. Prior to this, the only method of sending pictures was by despatch rider, train or ship.

Fig 2-3. A picture of President Armand Fallieres sent by cable from Berlin to Paris in 1907 using the Korn system. This system used a Selenium detector on the scanner and photographic film on the receiver.
From Sheldon & Grisewood 1929

However, transmitting grey-scale photographs over long distances caused problems. The signal degraded and transmission errors crept in causing artefacts to appear in the picture. The answer was to convert the continuous tone analogue pictures into digitally coded images.

Today, this appears anachronistic, even astounding. Surely digital picture coding is recent and needs complex digital hardware or even computer technology? Without computers but with ingenuity and simple valve electronics technology, a few such systems were developed and demonstrated. One system – the Bartlane system – became the primary system for news picture transmission up until the start of World War II.

Facsimile systems are now not limited to news pictures, weather charts or scanned flat images and documents. They include any capability of sending slowly scanned high-resolution images. Some of these images are the most impressive and revealing of modern times. Included in these are pictures from space probes of the surfaces of Venus and Mars, or pictures of the outer planets sent over hundreds of millions of miles. Arguably the most impressive distant image so far is a mosaic, taken from Voyager 1 from outside the solar system (Figure 2-4).

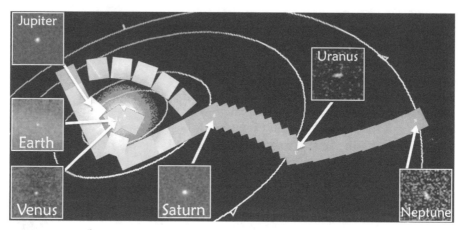

Fig 2-4. On 14th February 1990, Voyager 1 used its camera system for the last time to take a mosaic picture of our solar system from beyond the orbits of its planets. At a distance of 4 light-hours in this most distant vision of our home, the Earth appears as a pale blue dot. The Sun has shrunk to the size of Jupiter as seen from the Earth. Mercury is too close to the glare of the Sun to be visible and Mars and Pluto were not imaged.

From original courtesy of NASA/JPL

Moving Pictures – the Movies and Television

Movies happened before television solely because the technology was available first. However, that technology was quite basic. It involved only the chemical processes to make and develop the film and the optics and mechanics to transport the film through the camera and the projector. It needed electricity for the motor and light (but even that was not essential). Electronics only became essential when recorded sound accompanied the pictures.

Before television took hold, movies quenched the public's thirst for entertainment. The cinema had its major strengths: it was large screen (limited by the projector), high definition, eventually in colour, wide-screen and multi-channel sound. The technical quality was just so much better than television. Not only that, it became a great excuse for spending a night out, providing an escape from day-to-day troubles. It is hard for us to appreciate today how significant a hold the cinema had back when television first started. At that time, the cinema was 'king'.

Nowadays, the cinema is merely an alternative means of entertainment to television focused on the major benefit of showing *first-run* movies. With the advent of affordable 'home cinema' systems and movies on DVD, the cinema's technical edge is narrowing.

The cinema, or rather the technology for the cinema, is changing rapidly and radically. The technology that allows presentations from computer to be projected in a lecture theatre can easily present films from video. Indeed in small cinema theatres, video projectors fed by broadcast-quality videotape have already replaced the movie film projector. This technology is moving into home theatre systems. As more and more homes are equipped with such hardware, the threat to such cinema offerings grows.

Between late 1999 and early 2000, those people lucky enough to be able to watch 'Toy Story 2' in one of the specially equipped theatres in the United States and Europe were in for a visual treat. They experienced a total digital experience unlike any previous cinema presentation. Pixar Animation Studios created 'Toy Story 2' for the Walt Disney Company entirely in the digital domain within computing hardware. The movie was transferred to the distributed format (computer hard discs) with no intermediate film stage. The picture data was packed onto four disc drives using the state-of the-art wavelet compression techniques. Even with compression, the effective video data rate was 37 Megabits per second and the whole film occupied 32 Gigabytes. The innovation was not in the storage requirement: it was in the projection system. Texas Instruments had been developing its proprietary Digital Light Processing (DLP) technology for over two decades. The DLP Cinema projector used for 'Toy Story 2' was a prototype system that used three devices. Each of those devices held an array of 1,280 by 1,024 tiny 16 micron-wide steerable mirrors (see Figure 2-5). Fed directly by the data held on hard disc, the devices built up an image from over four million micro-mirrors adjusting their position to regulate light shining through. Though the technology is new, the method seems archaic and reminiscent of some of the wild imaginings of Victorian

Fig 2-5. The Digital Light Processing chip (left) – an array of 1,280 by 1,024 steerable mirrors. For colour projection, one chip is used for each of the three primary colours from which the colour image is optically combined. An ant's leg (right) shows the relative size of the mirrors.

Courtesy of Texas Instruments

inventors. Indeed, if twenty years ago, someone had suggested that we would be watching a cinema television system that encompassed four million mechanically scanned mirrors, they would probably have been ignored or ridiculed.

The picture from a DLP system falls short of the detail we see in a pristine 70 mm movie. Whether the DLP system or something like it will be the means to replace film in the long term is not important. What is important is that there are technologies that can perform in an alternative and, at times, superior fashion to existing traditional images on celluloid. The long-term danger to conventional film is real. Fundamentally, if it becomes more cost-effective to have digital movie presentations then capturing images on celluloid will in the future become outdated. Looking many years in the future, creating, distributing and even presenting 'digital movies' will undoubtedly be the way forward, whatever the means for doing it is. Copies of the master would give no degradation and multiple back-ups would preserve the data. All that would remain is to find a practical and completely reliable means of archiving tens of Gigabytes per movie. The sobering thought is that for movies on DVD that time is already here.

Cinema Television – 1930s Style

Bringing the technology for television into the cinema is not new. Though C. Francis Jenkins in the United States had a deep interest in television and moving picture technology, John Logie Baird in Britain had a focus on new technology for the cinema that was one of his more enduring legacies. In the late 1920s and throughout the 1930s he developed many systems for showing television to a cinema audience (see Figure 2-6).

After two years of development, Baird's 120-line colour projection television was demonstrated only twice in the Dominion Theatre in London's West End in 1938. This represented a massive leap forward from the sedate development of moving picture technology. Never before had colour television been demonstrated publicly, though Peter Goldmark, head of CBS Labs in the United States, was hard at work developing colour television based on the work done by Baird in 1928.[4] He gave the first US public demonstration on 3rd September 1940.

Although the three-colour *Technicolor* process had been developed in the 1930s, black-and-white movies were the exception rather than the rule right up until the 1950s.[5] The first full-length three-colour Technicolor film had been 'Becky Sharp' in 1935 – a screen version of 'Vanity Fair'. Large screen colour television projection in 1938 was breathtaking for those who saw it. The demonstrations led to further development and installation of

Fig 2-6. The Baird big screen television projector in the Marble Arch Pavilion, showing the 4.6m (15 feet) x 3.7m (12 feet) screen in position (1938–39).
Courtesy of the Royal Television Society ref RTS36-67

more advanced projection equipment, which were notably almost identical in principle to the video projectors of today.

We can see in Baird's cinema systems the early development of ideas that has led to widespread acceptance of television technology for showing first-run movies. Judging by the amount of effort he expended in it, Baird must have been deeply fascinated by the fusion of television with the movies. This stems from his work in 1929 on 'tele-talkies'; equipment for transmitting movies on television, known today as *telecine*. Though telecine is not something we would first associate with Baird, his legacy in this area can be traced through the incarnations of the Baird Company, through Cinema Television Ltd (CinTel) to Rank-Cintel, which became the foremost manufacturer of telecine equipment in Britain.

Baird is traditionally known far more for his achievements in developing and building a practical television system and delivering the message of the promise of television to an expectant public.

Television

The boundaries are blurring between television and the other methods of imaging. Television does however have one clear feature that discriminates it from all other imaging methods. Although Photography, the movies and facsimile meet most of the criteria for distant vision, none of them have *immediacy*. Television is the only distant vision system that really allows us to 'see at a distance' a live scene instantly. This was the dream of the inventors in the 19th century and before.

When they first appeared, photography, film and facsimile provided the basis for a wonderful new world of information and entertainment. Facsimile by cable or radio came closest to immediacy, but the scanning took tens of minutes just for one picture, and that picture had to be created in the first place. Back then, if television was to be like the movies, it would need to show several pictures a second and all of them captured, transmitted, received and displayed instantly. In the late 19th century, with development apace on photography and facsimile, television looked a long way off.

---

[1] WHEEN, F.: 'Television' (Century Publishing, 1985), p65

[2] BBC Horizon, 'Television is Dead, Long Live TV', 1996, director. Andrew Chitty

[3] BAIN, A.: 'Electric timepieces and Telegraphs', British Patent 9745, 27th Nov 1843

[4] UDELSON, J. H.: 'The Great Television Race' (University of Alabama), 1982, p152

[5] COE, B.: 'The History of Movie Photography' (Eastview Editions), 1981, p133

3   The Path to Television

*'The history of television forms one of the romances
of modern science.'*

A. Dinsdale, 'Television', 1926

Early Systems

Throughout the ages, and especially in the late Victorian and Edwardian periods, the various suggestions for television systems all displayed tremendous ingenuity and imagination. However, the *thought-pioneers* of that era based their solutions to television on technology that was existing or foreseeable. The biggest challenge was in the practical implementation of their ideas without the benefit of electronics. How could they convert a real-life scene to electricity, move the image from a remote place to the viewer using cable, or indeed the 'ether', and then reconstruct the image for viewing at a distant site – and all of this in an instant?

Some thought the answer lay in modelling what the eye and brain did. Our retina acts as a chemical camera, converting the image focused on the back of it to electrical signals. These signals are routed along the optic nerve into the visual cortex of the brain. There, a process indistinguishable from magic converts the signals into what we perceive as vision.

Mirroring Mother Nature became one of the dreams of the 19th century inventors, exercising imaginations, as amply described in the first ever brochure on the subject by Prof. Adriano de Paiva in 1880.[1] However, it became obvious that their ideas were just not practical. If we could first convert our scene into a flat area, like a painting or photograph, and then have photo-detectors arrayed evenly over the area, then there were two levels of unreality. First, it was not feasible at the time to have such a great number of photo-detectors (needing a minimum of several thousand for a reasonable picture). Second, it was considered next to impossible to have the signal from each photo-detector transmitted separately from its neighbours. The prospect of having several thousand detectors connected to several thousand pairs of wires across long distances simply failed the practicality test.

One Victorian inventor – a lawyer by profession – Constantin Senlecq of Ardres, France, in 1881, suggested a way round the problem of having so many wires, but still using an array of photo-detectors.[2] Only one connection should be made at any one time and that connection should be switched in sequence across every photo-detector in the array. This was similar to the scanning already practised in facsimile. A giant rotary commutator could have done the switching task. In proposing scanning, the inventors of the age hit more problems. First, they had to get the camera and the display in perfect step. Second, Selenium's slow response to light meant that scanning was going to be slow. Somehow the light value at each point had to be 'saved' until the next time the scanning returned to the same point. Third, building a large number of photo-detectors into an array was going to be a major challenge. In the absence of anything that would vaguely make this practical for television, such ideas fell into obscurity.

Ahead of its time

The idea of building a camera out of an array of detectors connected in parallel was valid but almost one hundred years ahead of its time. From the 1970's, solid-state cameras have used exactly this idea: a matrix of separate light detectors (not several thousand but several *hundred* thousand) are formed on the surface of a silicon chip. Each detector is separately 'wired' to its equivalent in a matrix of storage cells. As one picture is read (scanned) out of the storage array, the array of detectors integrates the electrical charge generated from the light of the camera's image focused on its surface. At the end of scanning out each picture, the exposed image is transferred instantly to the storage cell array and then read out in the same way as the previous one.

Scanning of some description seemed to be the answer. If an array of photo-detectors was out of the question, maybe just one high-speed photo-detector device could do the job. The device could move across the image or, more practically, the image could be moved across the device.

This was the principle of facsimile transmission. Selenium's slow response dictated that its movement across the document had to match its speed of reaction to those changes. This drove the scanning and transmission times for documents into tens of minutes. This was not thought to be a problem, as the alternative method of getting an image or document from one place to another was by stagecoach or ship.

Nipkow's Disc

In an interview given in 1933, Paul Gottlieb Nipkow (see Figure 3-1) described how he came across an efficient solution to scanning.[3] Fifty years before, in 1883, he had been loaned one of the brand new Bell telephones and was inspired by its simplicity. He focused on finding some means of building a simple yet practical system for 'distant vision'. On Christmas

Fig 3-1. Paul Gottlieb Nipkow, born 22nd August 1860 in Lauenburg, Germany. He was the inventor of the most popular method of opto-mechanical scanning for early television systems.

Courtesy of Andre Lange

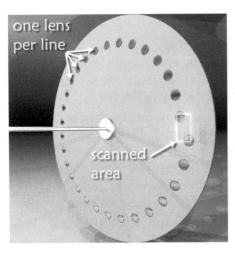

one lens per line

scanned area

Fig 3-2. The Nipkow disc configured as a single spiral 30-line camera.

Courtesy of the Author

Eve, 1883, the general solution to television, and specifically scanning, came to him. He was so certain that this was the right answer that he committed his own limited funds as a physics student in Berlin to apply for a patent.[4] His scanning method – a 'perforated spiral distribution disc' – was ingenious in its simplicity, yet the technology for the necessary high-speed photo-detector and signal amplification just was not there to make his system practical. With no immediate prospect of its use, he allowed the patent to lapse. Forty years after his patent was issued, the Nipkow disc was the method of choice for television scanning, used in some form or another by most of the early television pioneers.

The Nipkow disc needs some explanation as to how it works. The device is a flat disc with a series of apertures – lenses or pinholes – spaced equally around it, near to the edge (see Figure 3-2). The disc can be used for scanning either as part of a camera or as part of a display. The Nipkow disc used for display is the simplest part of the television system to describe. There are no lenses in the display disc, just pinholes.

Though each of the holes is placed at precisely equal angles around the disc, each successive hole is at a slightly different radius to the previous one. The step in radius is exactly equal for a very good reason. If we mask off a sector, we will see successive holes sweep by, the step in radius making them appear next to each other. Rapid spinning of the disc blurs to our eyes the movement of the pinholes and we see the path it traverses as a line. With the mask hiding the rest of the disc, it looks like there is only one pinhole sweeping by, forming successive lines stacked closely together. The most common version of the Nipkow disc uses one complete rotation to create one single television picture (see Figure 3-3). As the Nipkow disc is purely mechanical, different arrangements of the apertures could give almost any line scanning sequence and even generate multiple television images on each turn of the disc.

In a Nipkow disc television receiver, a light source is placed behind the disc and designed to cover the area swept out by all the holes. This is no ordinary light source as it must not only emit light evenly over an area that matches the television raster, but must be able to vary in brightness as rapidly as the video signal that is fed to it. The flat-plate neon was one such device, giving off a

Fig 3-3. An early 1926 off-screen photograph showing the effect of multiple images when viewing a Nipkow display disc.
Courtesy of the Royal Television Society RTS38-71

characteristic red light. The idea was developed early on in television's history. It was as far back as 1898, in the same year as the discovery of the element neon, that a Russian by the name of Volf'ke proposed using a gas discharge tube behind a Nipkow disc as a display device.

The Nipkow camera and the Nipkow display formed the core of early television systems. With both camera and display discs turning exactly in step – synchronised – the picture imaged on the camera would appear stable on the display. All that was needed then was a photo-detector, a signal amplifier and a means for getting the signal to the display system. Such was the beauty, simplicity and flexibility of the Nipkow disc.

This sounds easy today, but in the late 19th century, when all these ideas were surfacing in the wake of major engineering developments, there were two items missing. They were first, a fast-responding photo-detector and second, a means to boost the tiny electrical signal from such a detector. Television was going to have to wait until the development of electronics.

The Scots 'invent' Television

The absence of the technology did not stop the thought-experiments. The challenge of achieving practical television showed in the vast number of solutions proposed as being practical – if only they could be made to work. Most notable amongst these thought-pioneers was a prolific Scottish-born inventor who had already made his name in the field of X-ray photography and in cathode rays – the emerging electron-beam valve-based technologies. Troubled by negative comments from Shelford Bidwell on how to transmit high volumes of image information in synchronisation, Alan Archibald Campbell Swinton (see Figure 3-4) wrote a letter to the science journal, 'Nature', in June 1908.[5] He used the letter to outline an approach to the problem of scanning for television.

Fig 3-4. Alan A Campbell Swinton.
Courtesy of the Royal Television Society

His suggestion was to have two cathode-ray tubes – like those in an oscilloscope – with the beam in each deflected using the same waveforms to spread the spot across the screen. One tube had a fluorescent screen that would glow when the beam hit it. This was the display tube already patented and developed in 1897 by Karl Ferdinand Braun. The other tube used some as yet unknown material that would be sensitive to light yet somehow generate an electric current when the electric beam passed over it. Using the same generator to scan both the transmitter and the receiver ensured they were in synchronisation (see Figure 3-5).

In writing the letter, Swinton was recalling experiments that he had attempted in 1903–04. He had built an electronic camera tube and display system using the Braun cathode-ray tube along the lines of his 1908 proposal but had failed to get it to work. With hindsight, we can see that he had almost all of the elements of the eventual electronic solution. All he needed were the electronics to amplify the faint signal from his camera. When this experiment was repeated over thirty years later in EMI (Electrical and Musical Industries) Central Research Laboratories, but with the benefit of valve electronics, it worked![6]

His letter to 'Nature' in 1908 elicited no response. Three years later, at his presidential address to the Röentgen Society, he included, amongst other subjects, 'Distant Electric Vision' and revisited his earlier suggestion. No solution to this existed in 1911 and he freely admitted that his concept would not work 'without a great deal of experimentation and much modification.' He admitted that this was 'only an effort of my imagination.'

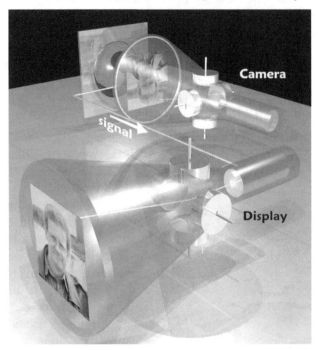

Fig 3-5. Swinton's concept of television by cathode ray tubes. At top, a lens focuses the image onto a photosensitive surface, which is scanned by the electron beam. The signal is fed directly to another tube (bottom), varying the intensity of its electron beam. Both beams are scanned by a common generator system, not shown for clarity.

Courtesy of the Author

At the time, the predictions in his 1911 address of which television was but a part, drew some interest. When Swinton revised his ideas and went into print in 1924, his three-part article was read avidly.[7] He placed a 'call to action' on corporate research laboratories where he believed the high funding necessary to develop such a television system could be met. His call to action may have been answered, as it appeared to spur on several of the main pioneers of television to commence or increase focus on television research.[8] Alexanderson at GE Schenectady, von Ardenne and Karolus in Germany, Takayanagi in Japan and Zworykin at Westinghouse all started research into television in the middle of 1924, a year after John Logie Baird.

What was Swinton's achievement? He had no working prototype, no demonstrable system. Swinton had only re-stated the essential elements of television – camera, display and their synchronisation – in terms of the Braun tube. There were other voices out there all proposing their particular view on the development of television. Could it be that his fame was caused by us looking back to find the closest prediction to the valve-based television age – an age that is now very much in the past?

For Swinton, that would be an entirely unfair view. Already a well-established scientist, his 1908 letter and many subsequent papers had direct influence on many corporate organisations. What Swinton had done was to clear away the dead-end ideas and focus directly on the areas that needed research and development and on the path that would lead to practical high definition television. We think today of Swinton as having great clarity of vision. His 'effort of the imagination' of 1908 remained the outline description of electronic valve-based television systems from the 1930s for the next 40 years.

Today though, our advanced digital technology solutions for television have gone beyond the predictions of Swinton. His time is now past. The concepts of the 19th century thought pioneers have now become far more relevant to today's television.

The Electronic and Mechanical Paths to Television

In the 1920s, the spread of practical valve electronics opened the way for two paths to television. One path used the newly developed valve electronics to provide the long-missing link, strapped onto essentially a solution from the previous century. Such a solution was based on scanning the image opto-mechanically (using Nipkow discs, mirror-drums and other quaint Victorian inventions). The other path saw the entire new field of electronics providing the complete solution to television. Very little existed, but the success of the Braun tube ensured its use as a display. Indeed, Kenjiro Takayanagi of Japan seems to have been the first to build a television system based on a Braun tube. Reminiscent of Farnsworth's 1927 demonstration of simple line, Takayanagi scanned a Japanese character, mounted on a mica plate and displayed the result on his version of a Braun tube. Though strictly not true television, Takayanagi's experiment was nonetheless a landmark event, though publicised some time after the event.[9] It was brave if not visionary to consider that electronics could hold the complete answer when most of the field of electronics had yet to be explored.

The true challenge for both approaches lay in making a practical tele-vision camera. The mechanical approach already had several methods for scanning and merely needed a suitable photo-detector and amplifier electronics. The electronic approach was starting from scratch. Consequently, it had a long way to go before being practical and usable in a broadcast studio environment. By the middle of the 1930s, the first electronic camera tubes were becoming practical to build in numbers and, with it, a whole new broadcast infrastructure to support electronic television. However, not everything was new: the techniques of live television programme making had been pioneered in the years before electronic television was available. Those programme makers had used the more technically crude mechanically scanned system to explore their artistic ideas and to learn how to adapt to the immediacy, and sheer openness of television.

Many countries developed mechanical television systems and used them to broadcast scheduled programmes. Almost anywhere in Europe in the early 1930s, tuning around the medium waveband (on the right day at the right time!), would result in television signals being picked up from Germany, Italy and Great Britain. With exceptional ionospheric conditions, signals from some of the many television stations broadcasting in the United States might just be heard.

After years of development, mechanically scanned television was highly developed. In the laboratory, demonstrations were given of high definition television – scanned using a mechanical approach. It is a popular misconception that the BBC's first television service had only 30-lines per picture because that was all a mechanical system could do. In reality, the only available transmitters and receivers in the early 1930s were on the medium and short wave bands, and a 30-line television signal was the maximum allowable number of lines you could transmit in one channel. By the early 1930s, high definition images of around 120 lines were being televised and displayed in the laboratory using mechanical systems.

The path to implementing practical television starts in a quiet town on the south coast of England where a Scotsman was recuperating from his last business venture – selling soap.

John Logie Baird

John Logie Baird's early exploits in trading in soap, together with his attempts at selling socks in Glasgow, safety pins to natives of Trinidad, producing jam in the jungle and re-selling honey in London, make him sound more of a commodity businessman than a great inventor. This all makes a great read in his autobiography, dictated to his secretary from memory many years later. When we read of these no less than hilarious exploits, we are, I believe, being treated to the traditional Scots way of self-

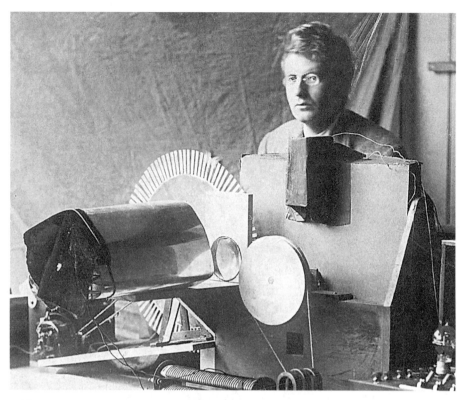

Fig 3-6. John Logie Baird, Hastings, 1923–24.
From original courtesy of Malcolm Baird, RTS36-02

deprecation touched with the dry humour of entertaining anecdotal adventures.

His recuperation at Hastings seems to have done him the world of good. It was here that he embarked on what was to be a life-long passion for television spurred on by one success after another and inspired by the realisation that he was the first of many to do so (see Figure 3-6). Within a short while, he had built the bones of a television system and achieved his first crude images. What is remarkable is that he achieved this by the end of 1923; very soon after electronic valves had become available and affordable to the public.

His first pictures were *shadowgraphs*. Shadowgraphs were relatively easy to produce when compared with true television. Shine a bright enough light onto the area being scanned, then stick in a hand, or a cutout of a face to partly block the light, casting a shadow over the detection area. The scanned detector would produce a binary image – black or white – of the silhouette. The light could be as strong as needed, even an arc-light if one could be arranged.

Baird used Selenium as his first photo-detector. It could only respond to changes in light slowly, creating blurred, smeared images lacking detail. This slow response forced him to use a slow picture rate, and very few lines. It was not long before he tried other photo-detection materials, though he kept the details to himself.

The picture quality, compared with what he later achieved was probably poor. Let us not get hung up on picture quality: the achievement in those early days was in scanning a shadow image and displaying the resultant pattern on a separate part of his system. However poor the picture was, Baird was the first to take these steps. Picture quality was something to strive for.

Fig 3-7. Bill Fox at his home in Golders Green in 1924. This photograph is contemporary with his reception of Baird's video transmissions from Hastings some 100 km away.
From original courtesy of the Royal Television Society RTS 36-22

Baird gave a demonstration to a journalist from the Daily News in January 1924. His Nipkow disc solution had four spirals per revolution of five holes each, giving a five-line shadowgraph image.[10] For display, he used ordinary filament torch-bulbs, one for each aperture in his camera disc. Despite the slow response of both Selenium and the torch-bulbs, his system was adequate to transmit and display simple shapes. The principle was there, though he managed to achieve all this with what Victor Mills, one of the Hastings radio amateurs, described as 'a pile of junk'.[11] The scanning discs were made from a tea chest from the local grocer with no lenses, just holes, in the camera disc made using the sharp point of a pair of scissors.[12]

Operating largely on his own, with friendly assistance from people like Mills, and without the benefit of a job to turn this into a funded hobby, Baird needed financial support to continue his work. From his past experience, he knew he had to 'sell' the idea. The best method that he knew was through publicity. He tried touting his message around Fleet Street but met only little interest. One Press Association journalist, W. C. Fox, was keen on developments in wireless technology. He assured Baird he would publish an article if he could hear the vision signal on the wireless. With friends and helpers in the Hastings Radio Society, Baird broadcast his shadowgraph television signal, which was duly picked up by Bill Fox at his home in Golders Green (see Figure 3-7).

Interest grew and enough funds came through for Baird to continue developing his approach. The torch-bulbs were replaced by a neon tube display. With the neon, Baird had the basis for a display that could respond fast enough and could reproduce sufficient brightness levels to display true television images. The aim was then to improve his camera system. Desperate for funds, Baird signed a contract with Wilfred L. Day, the owner of a radio shop in Lisle Street, in the Soho district of London, at a time when the region was more renowned for its trade in electronics than the trade of today.

The Day–Baird Letters

At auction in 1996, Will Day's grandson sold 76 recently discovered letters that had been exchanged between Day and Baird. The letters shed a little light on that largely unknown time in Baird's early television history whilst he was in lodgings at Hastings. In 1999, the new owner then resold the letters along with a 32-hole Nipkow disc attributed to Baird for £70,000. They now reside in the Hastings museum.

In one of these Day–Baird letters in June 1924, Baird mentions that he received a photocell, probably a US-made Case cell, from the supplier, C. F. Elwell of Kingsway, London.[13] A few months later, he mentions in a further letter that he was developing a *liquid* light-sensitive cell. Though its failure to appear in patents meant that it probably was not successful, it does show that Baird was willing to experiment as well as buying his components 'off-the-shelf'.[14]

Viewing in Reflected Light

The two weak links in Baird's television system were the photo-detector and the amplifier. These two areas had prevented television becoming a reality in the 19th century. By improving the signal conditioning from the photodetector and the video amplifier, he was able to 'see' objects in reflected light, and reported achieving this in June 1924.[15] He could not yet make out grey tones, but he could see the difference between light and dark in the scene (see Figure 3-8). Viewing in reflected light was a step in the right direction, sufficient to generate more interest and additional funds.

Fig 3-8. After shadowgraphs, Baird improved his system to show two-level images in reflected light. (1925)
Courtesy of the Author

In all this time, Baird was well informed of the developments elsewhere. In a letter to Will Day in September 1924, he refers to the work of A. Korn and Fournier D'Albe, and compares his results with those of C. F. Jenkins in the USA. Rather strangely, he admits that he could not say whether he or Jenkins was 'first' with transmitted shadows.[16]

The successes in Hastings led to an invitation by Gordon Selfridge to give demonstrations over a three-week period during April 1925 at his department store in Oxford Street, London. Money was tight and the £20 per week was a welcome fee. The demonstrations at Selfridges were simply a means of the store offering to the shoppers a glimpse of what the future may bring (see Figure 3-9). As it turned out, Baird's television was a great success, with huge crowds turning up to see this marvellous new development. Although the three weeks physically took a lot out of Baird, the subsequent interest and gifts of valves and batteries from a few companies made the venture seem worthwhile (see Figure 3-10).

Viewing in Light and Shade – Television at Last!

Baird recorded that his next achievement, resolving the subtle changes in brightness in the image, came on Friday, 2nd October 1925.[17] Though he had been able to see an image of sorts, reflected from crude cut-out faces and the like, this was the first time he had been able to see an object as it appeared in life. It had been a significant step to go from silhouettes to images in reflected light. It was an even greater step for his system to resolve the tonal range of light and shade that make up a scene.

Fig 3-9. The equipment and crowds at the demonstration in Selfridges store, London, 1925. On the left is a horn loudspeaker. Next to it is an amplifier with valves characteristic of 1924–25. The dark viewing tunnel ends in black cloth shrouding both sides of a Nipkow disc that is displaying images to the public. The light wooden box supports a 'chopper' disc. The box may house a Selenium photocell, as there are few wire connections.
Courtesy of Selfridges Ltd and the Royal Television Society RTS 36-86

The story goes that he ran down stairs to another office and fetched the office 'boy', 21-year-old William Edward Taynton. Taynton had for some time shown a keen interest in what Baird was working on. He had struck up a friendship with Baird that was to last for many years. In this, Baird showed a remarkable kindness to Taynton, recognising his partly disabled condition and ensuring the best for his friend.

Taynton became the first live subject for the improved system, allowing Baird to see what a human face looked like and to verify that his new development really worked. In asking Taynton to be the live subject, Baird later realised that his was 'the first face seen on television...'[18]

Despite Baird's assurance in print and in interview that Taynton was the first person to be televised (in full tonal range), there have been a few instances of people, many decades later, claiming that they were televised before Taynton. Robert Shaw of Larbert, Stirlingshire, recalled posing at the Temperance Café in Falkirk, where Baird was giving a demonstration.[19]

If this did predate Taynton's first appearance, a suitable explanation would be that Baird had not yet mastered television in full tonal range, and Mr Shaw was being televised in something like the binary level (soot and whitewash) quality of the Selfridges demonstration earlier on in 1925. The movement of Mr Shaw's face would have been quite impressive, regardless of lack of tonal range.

Fig 3-10. John Logie Baird, May 1925, Frith Street.
From original courtesy of the Royal Television Society RTS 36-06

The First Demonstration of Television

By January 1926, Baird felt ready to give a demonstration of television. He had met the requirements outlined in Hastings by Mr. Odhams of Odhams Press. He had said to Baird in early 1924,

> 'If you,' he said, 'could put a machine in the room next door and sit someone in front of it, and then on the screen of a machine in this room show his face – not a shadow but a face – then I am certain you would get all the money you want.' [20]

Though by no means an official Royal Institution function, some of its members made their way on the evening of 26th January 1926 to Baird's laboratory at Frith Street. Shepherded in a few at a time into the cramped

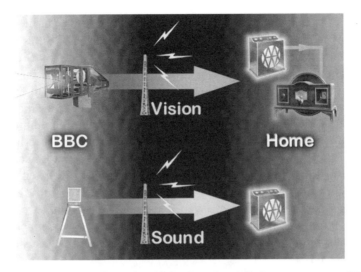

Fig 3-11. With the 30-line vision signal fitting in to the audio frequency range, all that was necessary to introduce a television service was a camera at the studio, an additional transmitter, an extra domestic radio in the home and a special display device – the *Televisor*. No new equipment or technology needed to be developed and the Televisor employed no active electronics.

Courtesy of the Author

demonstration area, they saw what is now regarded as the world's first demonstration of television, of a natural subject in reflected light with full tonal range. Today, we view that event as being a defining moment in the history of television.

A few months after the event, E. G. Stewart of the Gas, Light and Coke Company, filed a report on a follow-up meeting with Baird at Frith Street. The meeting took place with Baird during April 1926. Interestingly the objectives for television were viewed at that stage as being for a public service,

> 'in that pictures may be shewn of subjects in movement at the time of their occurrence, either in public, as in a Cinema, or in the home as an attachment supplementary to a broadcast receiver.' [21]

This last point shows Baird's thinking in 1926 and was what directed him towards offering a broadcast television service on 30-lines. He envisaged television as using the existing radio infrastructure, minimising development costs and costs to the consumer. All the consumer had to buy to get television would be a display attachment – the *Televisor* – that would ideally have a Baird brand name on the front, and a second radio to drive the Televisor (see Figure 3-11).

Stewart described the scanned area as being around 10 inches by

8 inches (25.4 cm by 20.3 cm) illuminated by 500 candlepower at a distance of 1 foot (30.5 cm). The scanning format used 32 lines (rather than the later broadcast format of 30) and the picture rate was 5 pictures per second. The displayed image was around 3¼ inches by 2½ inches (8.3 cm by 6.4 cm). Stewart reckoned that he was seeing a maximum frequency of around 1,600 cycles per second, probably worked out from the number of lines, picture rate and the detail along a line. This would amount to around 10 cycles per line; that is, without much detail. From his description of the refresh of the picture, line scanning was vertical and picture scanning was from left to right. Stewart was shown faces, but when he asked to see what the straight edge of a white card would look like, 'no facilities for introducing one's own subjects were existent.' Given what we know from the later Phonovision images and the Lafayette off-screen photographs, such a test would have made the geometric faults in Baird's system obvious (see Figure 3-12).

Fig 3-12. John Logie Baird looking down the viewing tunnel of an early television display at his premises in Frith Street, London. Part of a 'double-8' Nipkow lens disc can be seen behind him.

Courtesy of the Royal Television Society RTS 36-08

So here we have a picture format quite different to Baird's subsequent 30-line broadcast format, except for vertical line scanning. For comparison, the off-screen photograph of Oliver Hutchinson by Lafayette, the earliest photograph of a television image, first appeared in 'The Electrician' of June 1926 (see Figure 3-13). Hutchinson was Baird's business partner. It shows clearly a 30-line image with a vertical aspect narrower than Stewart reported and with fewer lines. Amongst a collection of Nipkow discs discovered in an old suitcase in Hastings was one with 32 holes and a 5:3 (vertical to horizontal) aspect ratio. This may have had something to do with Baird's system in early 1926; though what the disc was doing down in Hastings for all these years is anyone's guess.

Stewart's closing remarks show accurate objectivity.

> 'Mr Baird has solved in some measure the problem of transmitting a moving image via land line or ether over a considerable distance... at the present time, the image resulting is appreciably lacking in detail and so can have but little practical application.' [22]

Given the development possibilities emphasised by Baird and his partner Hutchinson, Stewart considered the system worthy of financial encouragement and development but warned off bringing the system to market without further develop-ment. He felt that the quality was simply not good enough and that facial shots alone were inadequate to bring the development to market. He believed that it would no doubt sell but only as a short-term novelty.

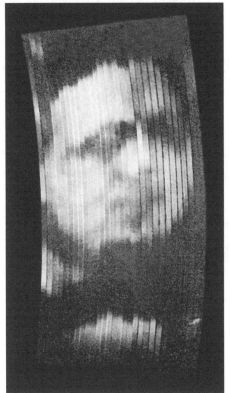

Fig 3-13. The earliest photograph of a television picture. This is Oliver Hutchinson, Baird's business partner. The white area at the bottom of the photograph is the next image of the top of Hutchinson's head. The picture therefore has a 3:2 aspect ratio (vertical to horizontal). This is a 30-line picture, showing that Baird had moved from 32-lines during 1926. The raggedness at the forehead is caused by faults in positioning the lenses and apertures around the circumference of the Nipkow camera and display discs – such as seen on the Phonovision discs. The vertical black lines on the right are caused by similar faults in radial position of the apertures of the display disc.
Courtesy of the NMPFT

Baird and his team had some work to do. They had to improve the picture quality, reduce distortion and most importantly, get the picture rate higher than 5 pictures per second. Such a rate is simply too low to see anything but simple, stable and slowly moving subjects.

Some weeks after televising Taynton, Baird moved to larger premises in Motograph House just a few streets away in Upper St Martin's Lane.

Baird's Television Format

There are some unusual characteristics of Baird's choice of format for his pictures. For all the experimentation in developing his format, he was always consistent in using vertical scanning. If the mask were set at either the top or bottom of the disc, the lines would sweep horizontally in a similar fashion to today's television. Such were the early systems developed in the United States and Germany. In Britain, Baird's system of television used a mask to the left or right of the disc giving lines that scanned vertically. After experimentation Baird settled on vertical line scanning, bottom-to-top, of a picture comprising 30-lines. In his television broadcast format from 1929 onwards, he chose the right-hand side of the disc (facing it) with the first line on the outside of the disc sweeping out the path of a TV picture from right to left. The Nipkow disc was turned at 750

Fig 3-14. A. F. Birch, 1931. The central photograph was a long exposure reputedly taken in the light of the flying spot scanner that formed the 30-line images either side. Despite the vertical streaks caused by the display scanning disc, these 30-line images show neither geometric nor frequency and phase faults. Given the quality at the time, and that the arc-scan is going the *wrong* way, this suggests the 30-line images above were possibly optical fabrications as described by Campbell and Birch in 'Television' September 1931. They demonstrate how good a 30-line image could be.
From original images courtesy of D. R. Campbell and R. M. Herbert

revs/minute giving 12½ pictures per second. The picture aspect was a vertical letterbox: a 'narrow-screen' television image, 7 units high by 3 units wide. Though the 30-line format was never officially accepted as a television 'standard', none of the engineering organisations challenged its specification prior to its use on the BBC Television service.

Though it seems unusual today to have a vertical letterbox and vertical scanning, Baird's reasoning was sound, practical and well thought out. The vertical letterbox format is near perfect for both the subject matter and the method of scanning. Baird realised that, with the limitations in transmitting television over a radio channel normally used for audio, he could only have an image that was low in detail. As the information content was at a premium, he needed to choose a format that made the most of what was in the scene and had the minimum of wastage. He matched the format to that of a single person view for announcer, singer, performer, actor or dancer (see Figure 3-14). From close-up shots of head-and-shoulders, to mid-shots and out to long-shots of full-length, the best image shape was a vertical letterbox. That shape would provide the least wasted area of the picture.

The vertical letterbox picture could be built up either of around 70 short

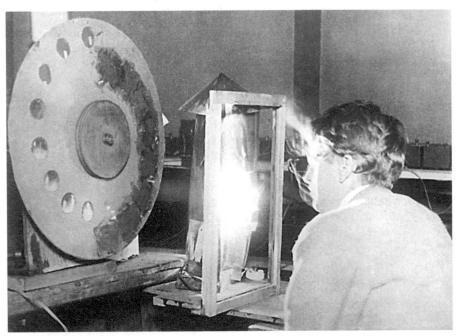

Fig 3-15. An early and unusual photograph from Frith Street laboratories in December 1924. The camera scanning disc has been half covered over, thus resembling the 'double-8' disc on display at the NMPFT in Bradford. Baird is facing a powerful lamp. The wooden frame may indicate the field of view of the camera.

Courtesy of R. M. Herbert

horizontal lines or 30 long vertical lines. The choice was straightforward. The fewer the lenses in a disc, the easier it would be to manufacture and the bigger the lenses could be in the disc. In addition, a horizontally scanned 7:3 aspect ratio picture would be distinctly keystone shaped and positioned at the top (or bottom) of the Nipkow disc. In the early days, when Baird was using the same Nipkow disc as a camera and display, he could image the scene on the right, say, and view the result on the left. This made for convenience.

There was an added convenience when Baird moved to the large scanning discs. Baird's early cameras used detectors that needed plenty of light to give a good signal (see Figure 3-15). Consequently he had to have large lenses in the Nipkow disc to capture as much light as possible. This required large diameter Nipkow discs to house the lenses. For those discs, some over 1 metre across, it was far more practical to form the image on the left or right side of the disc at the height of the motor driving it, than at the top or bottom. Positioning the object to be scanned crouched down on the floor or up near the ceiling would hardly be practical. Vertical scanning was simply the best all-round approach.

Baird described how he arrived at the format:

> 'The picture was made up of thirty strips. I had found this to be the minimum necessary to transmit a clearly recognisable image of the human face. To decide the shape of the picture most suitable to take in the face without waste space, I made endless measurements and ultimately decided on a long narrow picture in the ratio 7 high by 3 wide. The number of lines was arrived at by making drawings from photographs divided into strips. I tried experiments with different numbers of strips from 15 upwards, and came to the conclusion that 30 strips and a picture frequency of 12½ per second was the best compromise. The amount of detail which could be sent at that time was limited by the wireless transmitter. This also limited the number of pictures per second that could be sent out. It was a compromise between flicker and detail.' [23]

Non-linear Scanning

By the time Baird brought 30-line television to the public, he had modified his scanning method slightly. Whereas up to that point he had used equally spaced lines to build up the picture, he realised that medium and long shots of a single person would benefit from having a greater number of lines in the middle of the scanned area. Without adding lines to his format, he arranged for the centre group of lines to be bunched up and the outer lines to be spread out. This complicated construction of the Televisor display: it became essential to vary the width of the apertures on the Nipkow disc exactly in proportion.

Multiple Spirals – Multiple Images

A single spiral of lenses is just one solution to the Nipkow disc that Baird used in the late 1920s. A Nipkow disc based system can have as many solutions as the imagination allows. The holes or lenses in the disc determine where the lines occur that make up the picture. This means we can have quite elaborate paths for scanning the scene. As such, the Nipkow disc gives tremendous flexibility in building different television systems. By simply moving the mask around to the top or bottom of the disc, we can get scanning of the lines horizontally. By changing the number of lenses, we change the number of lines. The system described here has only one spiral per turn, giving one picture. Although this is convenient in terms of size, the single spiral disc is an inherently imbalanced design when the apertures are lenses.

We could just as easily have two successive spirals per revolution, which we might want to use to drop the disc speed or even balance the disc; both of which Baird did in his earliest experiments. With those two spirals, we can even offset them to provide interleaved or interlaced scanning for the primary purpose of reducing flicker. We might even want three successive spirals in just one revolution. If red, blue and green colour filters were placed over the three spirals, this would provide the basis for a colour television camera. We could even make a *stereoscopic* television camera if we had a single spiral on the outer portion of the disc and a similar one slightly further in with one photo-detector behind each of the spirals.

When we read today of some of the incredible achievements of Baird, of television in colour, stereoscopic television and interlaced scanning, all made decades before they were practical in the electronic camera, we need to appreciate that these were brought about by simple adaptations of the Nipkow disc. Once a Nipkow disc system had been developed, it just needed someone with creativity to get to these impressive and seemingly ingenious developments.

Of all these experiments and demonstrations of new television formats, by far the most successful was for colour television. Baird's experiments in 1928 used a 45-hole Nipkow disc comprising three spirals of 15 apertures. Unlike some of his other ideas, Baird was quite taken with the possibilities of colour television. The success encouraged him to develop a colour 120-line mechanical system in the late 1930s and a 600-line electronic colour camera tube in the 1940s.

His 30-line stereo experiments were not very successful. For these, he generated two images side-by-side and optically fused the images – one for the viewer's left eye, the other for the right eye. The main problem was that stereoscopy needs high resolution for realism. 30-line pictures can only show gross three-dimensional effects.

Seeing at a Long Distance

Baird's exploration into television continued right throughout the remainder of his life. What the public saw was a distinct but gradual progression from a laboratory exploration around 1926–28, where he was looking to show off new ways of implementing television (such as in colour, 3-D and infra-red) to demonstrations of how a practical broadcast television service could be achieved.

One critical aspect of practical television was establishing over what distance television could be achieved. There was no great understanding of what happened to pictures when distance degraded the signal. To that end, the first demonstration of long-distance television was across an existing telephone landline between two major cities. This demonstration was done jointly in the United States by AT&T (American Telephone and Telegraph) and Bell Labs on 7th April 1927 between New York and Washington D.C., a distance of over 250 miles. The images were 'reproduced with perfect fidelity' using a Nipkow disc based system with a 50-line picture at 18 pictures per second.[24] The cable link used three communications channels for transmission: one for the vision signal, one for timing pulses for the video and one for the audio signal.[25] This was a truly robust system but was only of use over telephone lines, where it is quite normal to use several lines simultaneously. However, taking up three precious broadcast frequencies for anything other than a demonstration would of course be wasteful.

The demonstration took Baird, his organisation and his underwriters by complete surprise.

> 'The AT&T were breaking our monopoly and taking from us our best talking point.... It was surprising that we had had so long a run (i.e. monopoly) as fourteen months'[26]

With Baird's video system running at audio frequencies, there was no technical reason why such a demonstration could not occur over existing telephone landlines to almost any distance. Baird staged a demonstration of television at a greater distance, just over 400 miles, from London to Glasgow, 'to revive enthusiasm after the damping effect of the AT&T's demonstrations.'[27] He sent one of his engineers, Ben Clapp, together with a display receiver off to Glasgow. Clapp set up the apparatus in the Central Hotel where images via landline from London were shown to local dignitaries. Linking up the two major cities was a great public relations coup that captured the imagination of his investors and the public with little or no further technical development. This was Tuesday 24th May, barely seven weeks after the US demonstration.

Clapp, his display and a few people were almost all that was required to stage the demonstration in Glasgow (see Figure 3-16). This was therefore not a demonstration of a technical breakthrough but one of current possibility. Baird could easily have done the demonstration months before. That he did it so soon after the US demonstration with just a handful of people was merely an attempt to upstage the Americans. He had to demonstrate at least to his financial backers that he was still the leader. However, the US achievement had a far superior image format containing almost two and a half times the information content of Baird's eventual 30-line television format.

Transatlantic Television

A short time after her husband, Ben, had left on a business trip to the

Fig 3-16. Ben Clapp at the Central Hotel in Glasgow posing with an early Televisor display that looks decidedly home-made. The demonstration to the Press took place on 14th May 1927. According to Mr Clapp's watch this photograph was taken at 8:14.
Courtesy of the Royal Television Society RTS 38-09

United States, the young and pretty Gwen Clapp started to receive male visitors, two at a time at her home in Coulsdon, Surrey. They came three nights a week at around midnight and left around three o'clock in the morning. The nocturnal goings-on may have attracted attention from the neighbours but the men were there to help with the equipment that Ben Clapp used as a radio amateur. Clapp owned one of the most powerful amateur radio transmitters in Britain. The high voltages involved required that, for safety reasons, the system be manned by at least two at any one time. A special license had been granted to allow the full power of 2,000 Watts to be used. This constrained the transmissions to a continuous two hours between midnight and dawn, to minimise any possible interference with national radio communications.

Clapp had set off with his display apparatus again – this time by ship to New York. He set up his equipment in Hartsdale, NY and, over the winter months of late 1927 and early 1928, tested reception of short-wave radio transmissions made (on 45 metres) from his home.

Baird kept almost no documentation on his experiments and demonstrations. However, Clapp kept every detail of his work. He kept the Phonovision discs used for the transmissions and the Logbook of his radio station, 2KZ.

Baird and Clapp were trying night after night to catch the right ionospheric conditions that would allow the television signal to be seen some 3,000 miles across the Atlantic. That radio path, covered regularly at the time by transmissions in speech, music and continuous wave Morse code, improved with the right ionospherics and at the right time of year.

On a few documented occasions, Baird used TV recordings – most likely the results of Phonovision experiments – as a source of material (see Figure 3-17). This bears some comment. If a Phonovision disc were played back, the video would be as we see today, grossly distorted and unwatchable with a picture rate of around four per second. The picture rate may not have been all that unusual as Baird used such low rates in demonstrations. Radio amateurs listening in would have heard the 'angry bees' sound of the television signal blasting out from Clapp's transmitter. This would have provided marketing 'pull' to television. Even if the amateurs had equipment to view the images, the distortion from Phonovision would have kept that small part of the nightly transmissions a mystery.

If Phonovision was so poor, why use it? It could just be that Ben Clapp was able to use the sound of the Phonovision video to judge the quality of reception. Baird would not have needed his laboratory equipment running and would not have required renting a landline from his laboratory in

London to the transmitter in Coulsdon. He could simply base himself at Clapp's house and use a gramophone to play the Phonovision discs over the air.

The New York–Washington D.C. and London–Glasgow television links in 1927 proved that landlines worked. However, these links were point-to-point communications just like a telephone call. The future of television very clearly lay in broadcasting rather than point-to-point communications. To take the step that would place Baird ahead of the pack meant that he had to make the demonstration using radio. This would have the greatest impact for the least effort.

G2KZ: "We have new record with face and hand."
24th November 1927

Fig 3-17. Entry in the logbook of Ben Clapp's amateur transmitting station, (G)2KZ.
Courtesy of R. M. Herbert

On 9th February 1928, under good radio propagation conditions, reception of the pictures was demonstrated at Hartsdale, NY. With Ben Clapp and Oliver Hutchinson of the Baird Company were Robert Hart, who hosted Clapp and the equipment, and an Associated Press news reporter.[28] The reporter described what he could see but it was little more than the shape of the head. He could see the mouth opening and closing and the change from face-on to profile. One of the subjects, Mrs Howe, he said, 'it was impossible, however, to distinguish her features.'[29] The ventriloquist's dummy head was reportedly much clearer. This is surprising, as the movement of the live subjects should have aided recognition. However, we may expect this effect when the picture update rate is very low. Any movement would then break up the image. It is likely Baird chose a low picture rate to minimise the effect of atmospherics. In any case, a picture of sorts was viewed. Clapp's memory of it was that there was little detail in the image, but that the shape of head and shoulders, and the movement came across well.[30]

The press went wild. The Manchester Guardian: 'Seeing Across the Atlantic'; the London Times: 'Transatlantic Television. Images transmitted to New York. Mr. Baird's success', and the New York Times: 'Transatlantic Television'. The New York Times editorial of 11th February hailed Baird's test transmission.

> 'His success deserves to rank with Marconi's sending of the letter "S" across the Atlantic – the first intelligible signal ever

transmitted from shore to shore in the development of trans-oceanic radio-telegraphy. ... Whatever may be the future of television, to Baird belongs the success of having been a leader in its early development.'[31]

From Clapp's original idea and Baird's innate drive for exploring his medium and staying ahead of the opposition, this was a victory of the imagination, an inspiration to others and source of wonder to the expectant public.

ET – phone Baird

The historic reception of pictures in Hartsdale, New York by Ben Clapp, assisted by Robert Hart, was the culmination of months of trying to get the right conditions for clear reception. Three nights a week for the winter months of 1927–28, Ben Clapp's powerful 2,000Watt short wave transmitter pushed out 30-line television images from his back-yard in Coulsdon in Surrey. The signals were not beamed directly at the United States. The 5-wire horizontal cage aerial sent signals out in all directions at a wavelength of 45m. With the right conditions, the signals could be heard all around the world, channelled by the ionosphere, to be picked up by the crude valve receiver in Robert Hart's home on the East Coast of the United States.

However, a fraction of the signal passed through the ionosphere into deep space. The rotation of the Earth swept the 'beam' across the heavens over the two hours of transmission, to be repeated regularly three nights out of seven over the months of the tests.

Carl Sagan in his classic novel 'Contact' had ET making contact with us through ET re-broadcasting back to Earth what Sagan had thought were the first television pictures – of the 1936 Berlin Olympics. Even if Sagan had known of Baird's transmissions, images of Hitler make a far better story than those of a ventriloquist's dummy. In reality, the narrow-band Baird 30-line transmissions have a greater range and a better chance of being detected. The periodic nature of the television signal and its very low bandwidth make this excellent material. ET should have an easy job of detection and decoding.

Dan Wertheimer, Chief Scientist of SETI@home and leader of the SERENDIP project at the University of Berkeley, has over twenty years of involvement with SETI observations. He provided me with the range at which we could detect the Baird transatlantic transmissions.[1] With our existing equipment – the radio telescope at Arecibo, Puerto Rico and the latest detector/analysers – we could detect the Baird transmissions from Clapp's transmitter out to one tenth of a light year without any difficulty. That is around one fortieth the distance to the nearest star other than the Sun – Proxima Centauri. If ET has a bigger antenna and better detection hardware, there is no telling how far out the signal could be detected. At the beginning of 2000, the signal had travelled 72 light years and was down to one half-millionth of the minimum detectable signal for our current technology, which is around 57dB below the power noise threshold.

Nevertheless, the SETI aficionados believe today that a continually repeating image is the best method of communication. It is conceivable that just the right radio receiving equipment could be pointed in just the right direction at just the right time, tuned to just the right frequency and operated by just the right aliens – ones without acid for blood!

A few weeks later, the New York press announced that radio amateurs in the Jamaica, Queens district of New York City had made a recording of Baird's television signal transmitted from England.[32,33] After interpreting the somewhat confusing report, it would seem that the amateurs picked up the signal, and recorded it onto a gramophone record. They then approached a piano tuner who told them the pitch of the television signal, and hence the line repetition rate. This allowed them to adapt their display equipment to show the 30-line picture and photograph the picture. They sent a copy of the photograph to London. There were no further reports, either from London or New York. Later attempts to track down the people and the claimed disc were unsuccessful.[34]

Demonstrations and Remonstrations

Much like the demonstration of television by landline from London to Glasgow in 1927, for Baird this feat reflected no tremendous engineering development. Baird handled his video signal exactly as if it were audio. He simply used amateur radio technology and his existing television equipment, relying totally on the right ionospheric conditions. The lesson from all these long-distance transmissions is that they were possible, easy to do (with the right conditions) and required very little customisation to what was by then existing technology.

Traversing the Atlantic was a journey of a week by ship and a risky pioneering venture by air. Just a few years before, in 1919, Alcock and Brown had made the first air crossing of the Atlantic in an RAF bomber – a Vickers Vimy. However, the thought of seeing moving pictures across the Atlantic, 'live' as they happened, thoroughly gripped the public's imagination – and Baird's.

There was however too much dependency on waiting for just the right ionospheric conditions to turn this into a practicality. After Clapp gave the demonstration in early February, he packed his bags and his 'Televisor' and set sail for home. On Clapp's return to London on board the S. S. Berengaria, he received images in mid-Atlantic on a machine that looked remarkably similar to that used a year before in receiving pictures in Glasgow.

Regular transatlantic television programming did not follow until the 1960s. Baird's demonstration was the pre-cursor to the new age, over 30 years away. The break-through was not so much in television but in communications technology. It took a great deal of 'rocket-science' (literally!) to make transatlantic television a financial and engineering practicality.

1928 – A New Scanning Method

In early 1928, the Baird Company rapidly grew in staff numbers. Amongst the many arrivals was 'Jake' Jacomb, who was appointed as Chief Engineer. Jacomb set about improving Baird's mechanical TV system in many ways. James D. Percy, Jacomb's personal assistant, recollected his style (see Figure 3-18);

'Infectious was his laughter and terrifying his rage, but no one who ever worked with this remarkable man could afterwards ever question his almost uncanny ability to find practical solutions to any technical problems, mechanical, optical or electronic with the limited range of tools and instruments available to him in his time.' [35]

Up until Jacomb arrived, Baird had been using a Nipkow disc with lenses to capture the reflected light from a floodlit subject. The light sensitive detectors were relatively large and insensitive, requiring huge lenses and hence huge Nipkow scanning discs to carry them. Percy recalls:

'The first of the vision machines, probably by virtue of its very size, is the one I will always most clearly remember. Installed (in 1928) in the Baird laboratories in Long Acre London, this basically consisted of a complex of shafting and scanning discs mounted on a framework of angle iron, and powered by a massive electric motor.'

Fig 3-18. Relaxing in the Copper-lined studio at the Baird premises in 133 Long Acre. From left to right, Livingston Hogg, Calkin (crouching), Jacomb (with saxophone) and Percy.

Courtesy of R. M. Herbert

'The largest of the scanning discs was made of wood, and was a full five feet in diameter. Around the disc's periphery were mounted a number of heavy glass lenses in the familiar Nipkow spiral formation and these, as the big disc wheeled around, appeared as a circular silvery blur. Later, when the disc was at rest, it became apparent that, should any of the lenses work loose, while the machine was in motion, it would leave its flimsy securing clips like a shell from a gun, and go straight through the walls, roof, or anyone standing in its path.'[36]

Percy later saw exactly this happen on a smaller scanning disc. The lens broke free from the Nipkow disc running at full speed and hit a lead acid battery causing an explosion.

Early in 1928, Jacomb introduced the flying spot system (see Figure 3-19), which dramatically simplified the hardware needs for television. The spotlight system reversed the role of cameras and lights. Instead of the Nipkow camera disc scanning the image over a light sensitive detector, the disc scanned a beam of intense light as a 30-line image over the subject in the studio. The floodlights were replaced by the photocells, with the studio in pitch darkness. The primary benefits were that a more efficient light

Fig 3-19. A Baird spotlight scanner at Witzleben in Germany, 1929. This system was mounted on a small trolley, which could conceivably have been moved whilst the image was being scanned. On the right, in the metal shroud, was a high output projector lamp. A condenser lens focused the light down the tapered fairing onto the surface of the aluminium Nipkow disc. The light shining through the apertures was projected onto the subject in a darkened studio.

Courtesy of R. M. Herbert

source, such as a carbon arc lamp, could be used, the large area photocells could now pick up all the light reflected from the subject and the Nipkow disc could be much smaller as it did not need lenses. A small Nipkow disc, with pinholes instead of lenses, could be made with greater precision than the huge lens discs, giving a better quality picture. Probably most important of all, is that the small aluminium disc could be spun at high speed, easily achieving 12½ pictures per second.

Fig 3-20. Findlay in the Long Acre laboratories, 1929.

Courtesy R. M. Herbert

Jacomb had transformed the way Baird's television was to develop, setting the scene for all future studio-based work using mechanical scanners. The only drawback was that the spotlight (or flying spot) system could not work in daylight. J. D. Percy recalls;

> 'Without any heat, in a cool dark studio and scanned by a single cold flickering flying spot of light alone, the subject could stay

any length of time in front of the scanner without suffering any discomfort. The only scanning disc required was a light aluminium one just 14 inches (36 cm) in diameter. Instead of the heavy and dangerous glass lenses that revolved with the original Baird scanning system, there was now only the single fixed lens that projected the flying light spot that came from the spiral of tiny apertures that was punched into the disc itself. And instead of using anything up to 10 or 20 kilowatts of lighting, all the illumination required was now provided by a single bulb, having a rating of under 1 kilowatt. Small wonder then that as this, the time of my arrival in John Baird's world of television, the spotlight scanner, although crude beyond belief to look at in its experimental form, was rightly regarded as a major step forward in the developing saga of the vision machine.' [37]

The timing is important, as it shows Baird had been using the floodlighting method until early 1928, and moved over to the flying spot method from then on.

From comments made in the Visitor's Book for the Long Acre laboratories, now held in the Royal Television Society, it would appear that several visitors in late 1928 and 1929 remarked on the tremendous improvement in picture quality. The spotlight system would have contributed to this in a major fashion: it would have given faster picture refresh rate and fewer picture defects through superior build quality.

The stage was set for those in the Baird Company with the entrepreneurial spirit to promote television as a public service. Though engineers like Percy did not believe the system was ready for the public, the company had to generate revenue to cover the massive costs in its development of a practical television system (see Figure 3-20). These costs would continue to rise, eventually, by 1931, crippling the Baird Company.

---

[1] DE PAIVA, A.: 'La téléscopie électrique, basée sur l'emploi du sélénium' (published in Porto by Antonio Jose da Silva), 1880, 48 pages

[2] SENLECQ, C.: 'The Telectroscope of M Senlecq, of Ardres', *The Electrician*, 6, 5th Feb 1881, pp141–142

[3] *New York Times*, 6th Aug 1933

[4] NIPKOW, P. G.: 'Elektrisches Teleskop.' German Patent 30105, 15th Jan 1885

[5] SWINTON, A. A. CAMPBELL: 'Distant Electric Vision', *Nature*, 78, 18th June 1908, p151

[6] BRIDGEWATER, T. H.: 'A A Campbell Swinton FRS' (Royal Television Society Monograph series), 1982, p19

[7] SWINTON, A. A. CAMPBELL: 'The Possibilities of Television with Wire and Wireless', *Wireless World and Radio Review*, 14, three articles: pp51, 82, 114

[8] ABRAMSON, A.: 'Zworykin, Pioneer of Television' (Univ. of Illinois Press), 1995, p48

[9] ABRAMSON, A.: 'The History of Television, 1880 to 1941' (McFarland & Co.), 1987, p94

[10] HERBERT, R. M.: 'Seeing by Wireless' (R. M. Herbert), 1st edn, 1996, p5

[11] MILLS, V.: Interview, *Television* (Granada Television), 1985

[12] HERBERT, R. M.: *ibid*

[13] HERBERT, R. M.: Private communication with the author, Jan 2000

[14] BAIRD, J. L.: Letter to W. L. Day, 13th Oct 1924

[15] BAIRD, J. L.: Letter to W. L. Day, 16th June 1924

[16] BAIRD, J. L.: Letter to W. L. Day, 27th Sep 1924

[17] BAIRD, J. L.: 'Sermons Soap and Television' (Royal Television Society), 1988, p57

[18] BAIRD, J. L.: 'Sermons Soap and Television' (Royal Television Society), 1988, p57

[19] *Falkirk Herald*, 24th Sep 1992

[20] BAIRD, J. L.: 'Sermons Soap and Television' (Royal Television Society), 1988, p45

[21] STEWART, E. G.: Report, Gas, Light and Coke Company, Apr 1926

[22] STEWART, E. G.: *ibid*, Apr 1926

[23] BAIRD, J. L.: 'Sermons Soap and Television' (Royal Television Society), 1988, p62

[24] 'Far-off speakers seen as well as heard in a test of television', *New York Times*, 8th Apr 1927

[25] Bell Telephone Laboratories paper presented at Summer Convention of the AIEE, Detroit, Michigan, June 20–25, 1927

[26] BAIRD, J. L.: 'Sermons Soap and Television' (Royal Television Society), 1988, p74

[27] BAIRD, J. L.: 'Sermons Soap and Television' (Royal Television Society), 1988, p76

[28] HERBERT, R. M.: 'Seeing by Wireless' (R. M. Herbert), 1st edn, 1996, p7

[29] 'What America Saw', *Manchester Guardian*, 10th Feb 1928

[30] CLAPP, B.: Private communication with the author, Sep 1982

[31] Editorial, *New York Times*, 11th Feb 1928

[32] 'Radio amateurs here catch London picture', *New York Times*, **19**, 14th Mar 1928, pp3,4

[33] 'Television Record hurried from Jamaica to London to disprove cry of "fake" ', *Brooklyn Daily Times*, 14th Mar 1928

[34] SHIERS, G.: Private communication with T. H. Bridgewater, 5th Oct 1982

[35] PERCY, J. D.: 'The Vision Machine' (Unpublished memoirs), 1979, p25

[36] PERCY, J. D.: *ibid,* p19

[37] PERCY, J. D.: *ibid,* p28

4 Phonovision

*'Phonovision – the undoubted father of all
video recording systems'*

J. D. Percy, 'The Vision Machine', 1979

The BBC disc

In 1966, as part of the 30th anniversary of the start of 405-line television, research work unveiled a disc held in the BBC Sound Archives. This contained a recording in the 30-line Baird video format. BBC engineers used the best equipment available to get pictures off the disc. With the benefit of audio filter banks, phase correction equipment and special hardware to drive an electron tube display, they managed to get several photographs of single pictures from the discs. The results were good and caused great excitement.

Thornton H. Bridgewater OBE was extremely interested in the results. He had joined the Baird Company in 1928 and transitioned to the BBC when it started its 30-line service in 1932. He was one of the BBC's three television engineers – Birkinshaw, Bridgewater and Campbell (coincidentally with initials B.B.C.) – who ran the BBC's 30-line television studio between 1932 and 1935. He worked in BBC Television Engineering, primarily in Outside Broadcasts, until he retired in late 1968 as Chief Engineer, BBC Television.

If the contents of this archive recording were indeed of 30-line television from the BBC period, Bridgewater would have been one of those directly responsible for the corresponding programme.

It eventually became clear that this disc was a late recording of 30-line television, intended as a test disc rather than of broadcast material. Not only that, it contained exactly the same material as a disc sold through Selfridges in 1935 by the Major Radiovision company under R. O. Hughes, with F. Plew as General Manager and Technical Adviser.

Hints of Phonovision

It became clear that not all 30-line recordings were attributable to Baird and not all were from his very early experiments in video recording, as suggested by his patents. The Major Radiovision disc played back a Baird format video signal from a standard 78 rpm turntable. There was another, quite different type of 30-line disc. This contained what sounded like a slowed-down version of Baird's video signal. Unlike the Major Radiovision discs, this type of recording had defied all attempts to get pictures back from it. In an internal BBC memo dated 1981, two such recordings were referenced: 'Woman Smoking', 1928 and Geoffrey Parr's disc (belonging to the Television Society) also dated 1928.[1] Neither had been successfully played back.

> 'In September 1979, Tim Voore reported that there was no H.F. (high frequencies), and only blurred images – nothing distinct.'

The memo's author goes on to pose,

> 'I have not compared the two signals (from the two separate discs). I wonder if they are the same ... or at least the same standard?'

The memo refers back to earlier correspondence that Geoffrey Parr, the holder of one of these supposed *Phonovision* discs, had entered into with an enthusiast. In this 1960 letter, Parr referred to the recording as, 'one which J. L. Baird owned and which he gave to me just before his death'.[2] Parr had played a major part in the early development of television and was, in 1960, Honorary Secretary of the Television Society. Parr continued, 'I shall be most interested to hear what results you obtain from the recording as I have never seen the picture myself!' From the enthusiast's reply a few weeks later, no pictures had been seen. However, Parr confirmed that both he and Tony Bridgewater thought that the signal was 'a little low-pitched'. He went on to say, 'However, there is no doubt that the record was made under the supervision of J. L. Baird and the only thing that I can think of is that it may have been recorded at a higher speed – although this is very unlikely.'[3]

By the early 1980s, this was all that was known about Phonovision and the television recordings of the 30-line era.

The Search begins

The BBC used short extracts from their archive disc on an audio documentary on Long Play (LP) gramophone disc made to celebrate the 40th anniversary of the start of the 405-line television service in 1936.[4] These were the very extracts that I had discovered in 1981 and that had started the work described in this book. Of course, at the time, I had no idea from where the extracts originated – but the BBC had to have access to them to make the documentary.

Tantalised and inspired by those images, I began to search for more material. It became obvious early on that these discs were in no way common. The expectation though was that if there was one recording, there must be a great deal more. For all anyone knew, the BBC could well have had a whole library of such early 30-line video recordings. Not surprisingly, the first stop was at the BBC. My initial direct approach through Pat Leggatt, the Head of BBC Engineering Information Department, yielded no information on 30-line recordings.[5] With the Baird Company no longer around to approach, there were only a few avenues open to me. I came across one possibility: a group dedicated to using mechanically scanned low definition television for amateur radio.

Fig 4-1. One of the NBTVA's members, Deryck Aldridge, with a home-made 32-line television camera employing drum scanning.

Courtesy of the Author

The NBTVA

Though amateur radio supports image transmission, the image formats mostly involve sending still pictures. The closest general imaging format was called Slow-Scan Television (SSTV). However, with a minimum scan time of eight seconds per picture, and each picture being normally different from the next, the term 'television' is misused. This is more of a 'fast fax'.

A few radio amateurs wanted to develop a system for transmitting and receiving true moving images rather than SSTV. In April 1975, they formed a club based in Nottingham, England dedicated to developing techniques that would bring television to the radio amateur bands (see Figure 4-1). Forced to fit a television signal into a signal bandwidth only suitable for speech, the amateurs came up with a solution from television history. They resurrected mechanically scanned low definition television as most suitable for what they needed. Not only would the television pictures be low in bandwidth, the equipment that both generated and displayed the television pictures could be built incredibly cheaply. The television camera and display comprised lenses or mirrors, discs, motors and the simplest of electronics. They later named their group 'the Narrow Bandwidth Television Association' (NBTVA), headed up by a retired teacher, Doug Pitt. Pitt's experience with mechanical television stemmed back to the days when, as a schoolboy in the 1930s, he had built a receiver to watch BBC Television on 30-lines.

The NBTVA seemed to be the most likely place to know if any such recordings still existed. After an initial approach in early 1982, Pitt sent a short reel of audiotape containing what was thought to be 30-line recordings made at the time of Baird. He was enthusiastic about what the computer processing might achieve.

The tape contained three recordings, one of which was supposedly of a woman smoking a cigarette (see Figure 4.2). Despite my rudimentary processing, the woman was difficult to make out, though there was a white line apparently hanging down from her mouth. Could this be the cigarette?

Video is fickle about being recorded on tape – especially when, like these 30-line recordings, it was recorded directly without processing. It could well have been that the copying process had distorted the video signal. I set about trying to find the original discs to transcribe them in a controlled fashion.

Fig 4-2. An early uncorrected image.
Courtesy of Author

Ray Herbert

A short news item on the 30-line image processing appeared in the NBTVA Newsletter. Ray Herbert, a retired engineer who started out in the Baird Company, spotted this item and made contact. Herbert's close friend, Ben Clapp, had a video recording given to him by John Logie Baird before the Second World War. In fact, Baird had given him several, but a German bomb on Clapp's house during the War destroyed all but one.

Fig 4-3. Label from Ben Clapp's Phonovision disc. The date on this label is 20th September 1927, making it the world's earliest known recording of a television signal.

Courtesy of R. M. Herbert

Clapp's link with Baird went back to the start of the Baird Television Company, then called 'Television Ltd'. Clapp had been Baird's first assistant and played an important part in Baird's early television experiments. In 1982, Clapp was a bundle of energy and had the timeless air of an enthusiastic engineer caught up long ago in a world-class development. He had slimmed down over the years but this was unmistakably the same man in the photographs receiving television by land-line from London in Glasgow's Central Hotel, and later receiving the first historic television pictures in New York, transmitted by Baird in England.

Despite this, his memory of events in those early days was disappointing and unfortunately, that included Baird's video recording experiments. This after all was not something with which he had been directly involved.

In early summer, 1982, I met with Clapp to transcribe his sole surviving Phonovision disc (see Figure 4-3). The date on the label was a surprise – 20th September 1927. The disc held images almost 55 years old. The label identified this as an experimental test record, strongly suggesting from the date and from Clapp's close association with Baird that Baird himself might have been directly involved. This was the first Phonovision disc I had come across and it had been made just a year and a half after Baird's historic demonstration of television. Could this really be genuine? Why had it not been heard of before?

The answer came partly from Clapp himself. He vaguely recalled

Pictures and Frames

This book refers to 'television image', 'picture' and 'frame'. Whereas 'image' and 'picture' refer to what we see, 'frame' has specific technical meaning.

A 'frame' is a single picture comprising a complete set of lines. In the case of Baird's format, we talk of 30-line frames displayed at 12½ frames per second. There is a close analogy with cinematography, where the definition of a frame is one of a succession of images that together comprise the motion picture.

that the recording was faulty and that the image played in the laboratory was impossible to recognise. His wife, Gwen, thought the image looked like a cabbage.

Ben Clapp's disc, the first to be restored and to date the earliest known, had a strange radial pattern on the recorded disc surface. The pattern was caused by the tiny deviations in the groove for each television picture lining up on each turn of the disc. There were, within a very small error, exactly three 30-line frames – 90 lines – per revolution of the disc. Consequently playing back the disc at 78 rpm yielded a slow frame rate of around 4 per second. The disc had a label identifying it as a Columbia Graphophone Company Test Record with the date stamped on it. Written on the disc surface was the reference, SWT515-4.

Drawing a Blank

The search for more discs spread to museums and television institutions in Britain. They were asked if they had any early video recordings on disc from the 30-line period or had any information relating to them. In July 1982, Keith Geddes, curator for television at the Science Museum in South Kensington replied, 'It ought to have occurred to me that the pictures obtained from low-definition television recordings were susceptible to improvement by computer techniques, but I must confess it didn't.'[6] He knew of no other recordings that survived but he recalled being loaned, 'at the time of the 1980 exhibition, a disc autographed by Baird that was alleged to have contained a picture of a woman smoking a cigarette, but all attempts to get anything recognizable off it failed…' He had the opinion that, 'one or two commercial discs of still pictures are all that survive'.

He referred me to Tim Voore, a BBC engineer who had helped him out. Voore had been instrumental in getting some still pictures from the 'commercial' discs whilst at the BBC in the 1960s. Voore had kept in close communication with Tony Bridgewater. When Bridgewater heard of the work, he took a personal interest in the recovery of the 30-line television pictures, and used his influence and the respect he had in the history of television to try to find out more information. Meanwhile, Tim Voore had managed to locate the only disc in the BBC containing a recording of 30-line television. This was a disc whose label declared it a 'copy of a recording of an actual transmission'. Disappointingly, the BBC disc turned out to be a transcription, and only a fair one at that, of the commonly available commercial 30-line test disc from the Major Radiovision company.

Tony Bridgewater described his own disappointment:
    'The BBC, including myself at the time, was well and truly

spoofed by the Major Radiovision "find" in the 1960s. Never having then heard of MR we all assumed that <u>any</u> 30-line recording must have emanated from Baird. And as you know IBA were taken in too...'[7].

Bridgewater was referring to the Independent Broadcasting Authority (IBA) who, from their work in 1971 right up until their museum – the IBA Gallery – was closed down, maintained that their Major Radiovision recording *was* Baird Phonovision.

The BBC was told the bad news – the only 30-line disc they had was merely a copy of the common Major Radiovision disc.[8] The original Major Radiovision disc had been intended purely as a source of test signals for use in lining up displays. There is no record of the video signal on it ever being transmitted.

Bridgewater had gathered just a few items on Phonovision and knew of two Phonovision discs: one at the Royal Television Society and the other in private hands.

John Ive owned what turned out to be probably the most interesting of all the Phonovision discs. I visited Ive's office in Basingstoke, Hampshire one Sunday in July 1983 to transcribe his disc. Mostly similar to Ben Clapp's disc, there was one major difference: the radial structure that weaved across the other discs formed perfectly straight spokes radiating out from the centre of the disc. This structure meant that all the timing for each picture frame lined up as near perfect as a ruler laid across the lines could tell.

This disc bore a hand-written gummed label signed 'J. L. Baird' with the words, 'Baird Phonovision Record Made 28 March 1928. Shows Lady moving head and smoking cigarette.' Embedded on the disc surface is the reference 'RWT115-3') and the name 'Miss Pounsford'. This was the original disc of which Doug Pitt had supplied a copy and which had been loaned to the Science Museum in 1980.

The Royal Television Society

At the end of Baird's British Association lecture on television's potential at Leeds in 1927, W. G. W. Mitchell proposed the customary vote of thanks. Baird's demonstrations had made a deep impression on the whole audience, inspiring Mitchell to make a call for action. Mitchell suggested that 'in view of the wide public interest in television' a society be formed to further the development of television, Noctovision, Phonovision and allied subjects (all of which had been discussed or demonstrated that day) and to give a stimulus to this 'new branch of science'.[9]

The Television Society was formed, later to achieve 'Royal' status, and rapidly became the centre for developments in television in Britain. Today, the Society has moved on from its engineering heritage, yet it still maintains a small archive of historic material. One of these items is a disc, donated by Baird and sporting the hand-written label,

'Baird Phonovision Record, Made in 1928.
Shows Man's head in motion'.

Fig 4-4. The Phonovision disc belonging to the Royal Television Society. RWT620-11 10th January 1928.

Courtesy of the Author

In January 1985, the Royal Television Society loaned me their precious Phonovision disc for transcription. Although the label was hand-written, the disc surface bore the characteristic reference number. In this case, it was RWT620-11) (see Figure 4-4).

The quality of this recording was remarkably good, distinctly better than Ben Clapp's 1927 disc. The image was far more stable and there was a good deal of movement.

The Baird Company Network

Ray Herbert had for many years maintained a strong network of retired Baird Company engineers. Shortly after our meeting in mid-1982, he exercised his network to see if anyone else other than Ben Clapp had early television recordings on disc. His search revealed that there were two other discs – one owned by E. G. O. Anderson and another by H. C. Spencer.

Spencer had been employed at EMI and discovered by chance his Phonovision disc in a pile of rubbish scheduled for disposal. The reference number was RWT620-4, with the label date-stamped 10th January 1928.)

Anderson had been Baird's assistant throughout the war years and had been given the disc by Baird possibly as a keepsake. Anderson's disc turned out to be interesting but for different reasons. The disc was identical, apart from the label, to the Royal Television Society's disc. This was another pressing of RWT620-11) but now with a conventional Columbia Graphophone Test Record label dated 10th January 1928. The first block of letters and numbers possibly identified a particular recording session on one date and the '-11' indicated a 'take' number for that session.

By 1985, almost all Baird's contacts and work colleagues had been approached, thereby exhausting all obvious sources of Phonovision discs. With almost every new contact, there was someone who believed they had a priceless Phonovision disc when in fact it was yet another copy of the commercial Major Radiovision disc. This disc was sold through Selfridge's in 1935 at the end of the 30-line era. It held a collection of test stills, supposedly for providing a line-up picture for 30-line television displays. This was a conventional two-sided 78 rpm disc distinguishable by its unique red label.

Given the experimental nature of Phonovision, it was not surprising that the people who had copies either worked for Baird or would have received one as an exhibit for a museum. What was puzzling was that the Science Museum did not appear to have one. Baird had donated his first experimental TV apparatus to the Museum and ought to have included a Phonovision disc or two.

Final Phonovision?

Unexpectedly, in late 1986, Eryl Davies at the Science Museum in South Kensington made contact. He had accidentally come across a Phonovision disc stored between thick layers of cardboard in a filing cabinet, together with the original exhibit tag. The disc had been donated in 1935 and the exhibit tag described the disc.

"PHONOVISION" RECORD, 1928
Lent by
BAIRD TELEVISION, LTD.

"There is no reason why television signals should not be recorded and used to reproduce the televised signal at a later date, although the cinema film forms an easier method than electrical storage at the present time. This gramophone record, made by Mr. J. L. Baird in 1928, is an example of an early attempt to make an electrical, as opposed to an optical record of a scene.

The record, when played with an electrical pick-up, the output of which is connected to a television receiver of appropriate design, reproduces a picture of the scene recorded. The television signal currents are made to actuate a recording needle in cutting the record and the technique is quite straightforward within the frequency response limits available. As the limit is probably, in this case, something below 4,000 cycles, the pictures reproduced can only be small and very crude. Something of the order of half a million cycles would be required for good definition. The record can therefore be regarded as a purely experimental device demonstrating the principle, and as such was successful in that the results, though of poor quality, were recognisable."[10]

This Phonovision disc had 3 pictures per rotation just like the others. It had lost its label, but on the disc surface was the reference RWT620-6.) The discovery of this disc came at a time when I had studied and processed all the other five Phonovision discs thoroughly and had already published a detailed article on them.[11] The inventory tag provided further useful evidence of the disc's authenticity. The reference number and the reference to 'Wally' on the disc surface strongly suggested that this had also been recorded on 10th January 1928. Of the six Phonovision discs, four came from this one recording session and two of the four were pressings of the same take. From 1986 to the time of publication of this book, no further discs from Baird's experimental period have been found.

The distinctive feature of these discs is that they are all single-sided pressings made on professional equipment. That equipment had only become available in 1927, just prior to the dates on the discs. Baird had used the latest in electrical recording technology for his experiment. What then becomes perplexing is that, despite the professional and state-of-the-art nature of the discs, the recorded signal has been so distorted during the recording as to make the material on the discs unrecognisable.

Absence of Evidence

What we know about Baird's achievements comes from several sources: his patents, the write-ups in the press, hobbyist periodicals of announcements and demonstrations, and some photographs.

In researching historic developments, photographs are not a reliable source of material – particularly when associated with Baird. Tony Bridgewater explains,

> 'One of the problems common to many pictures of Long Acre *(post-1928 – Bridgewater's time with Baird)* activities is the fact that photographs in the press were often less than strictly authentic. Sometimes so as not to give away 'vital secrets' – other times simply to move apparatus around to facilitate the photography in, as likely, a small room.'[12]

Press reports need 'calibration' – especially if authored by a non-specialist. All too often today we see wild inaccuracies caused by ill-prepared journalism tackling the pressures of the deadline and the need to get 'a story'. Coupled with that, differences of opinion amongst journalists produce different versions of the same story. The newspaper is safer when treated as an expression of opinion, not absolute incontrovertible fact.

The most appropriate example to this book appeared in the London Times of 28th December 1996. The newspaper carried a half-page article on the restoration of the earliest-known recording of broadcast television. It described how, 'computer experts, working for the Museum of Photography, Film and Television in Bradford, have decoded the discs to reveal the first recordings of moving television pictures'.[13] They referred to the subject as being 'Emily Pounsford' (should be Mabel) and the dummy head as 'Stooky Bob' (should be Bill). Two days later, the Daily Mail covered the story, again with a half-page spread.[14] The restoration had grown to become a team effort. 'Now 70 computer experts at the Museum ... have managed to decode the discs which reveal the first recordings of moving television pictures, a breakthrough that has thrilled historians.' As this book reveals, the reality is that the author restored the discs, working on his own with no connection with the Museum other than providing it with material to exhibit.

The media today seem only to be able to work in 'sound-bites'. Rather than tell a story straight, the media will focus on something that will grab the attention of the viewer or listener who would not normally be interested. Bill Fox was interviewed in 1983 for a documentary film. After a full day of filming the 95-year-old frail Fox, the documentary makers used only one sentence referring to John Baird, 'And he said very quietly to me that I have come to tell you that I have achieved Television'.[15]

John Logie Baird has had the privilege of having several books and monographs written about him and his life. The earliest, published in 1933, ought to have been the most accurate. The author, Ronald Tiltman, presented a romanticised version of events, focusing on the man, and describing the headline impact of his achievements, rather than any details. Baird's business partner/manager, Moseley, wrote another account in the 1950s. This captured Moseley's sentiment that Baird had been hard done by – especially by the BBC – and had not achieved true recognition. Both Tiltman's and Moseley's accounts of Baird are personal and emotional and not objective or indeed reliable.

Baird's own story lay unpublished for over 40 years after his death, although it provided the raw material for Moseley's and later Margaret Baird's biographies. When the Royal Television Society published the autobiography in 1988, it appeared as a breath of fresh air. There is a dry Scots humour that pervades his story. Baird is objective, at times proud of what he achieved, astonished at what he had done and self-deprecating about his failures both in business and in his technical development.

The most reliable documents that we have on Baird and his works are probably the patents. Patents do need to be treated with some caution, though; they did not necessarily describe a working system (and indeed usually included deliberate errors to foil the competition). When taken together, the patents, biographies, photographs, literature and living memory are all the evidence we have.

Equipment

The most notable piece of original Baird equipment is apparatus that is part of a prototype 16-lens television system. Sometimes called the *double-8* system due to its two spirals of 8 lenses each, the better known of the two surviving instances is on display at the National Museum of Photography, Film and Television (NMPFT) in Bradford. Baird had donated this in October 1926 to the Science Museum in South Kensington along with a ventriloquist's dummy head – a test subject from those early days. The prototype appears to be a model of the apparatus that Baird used, rebuilt for the purposes of offering it to the Museum. The model is incomplete, supposedly for reasons of preserving his commercial security.

Baird's 1925 shadow-graph system

This had a distinctive 'double-8' Nipkow disc – two 8-aperture spirals on one turn of the Nipkow disc. The double spiral, though unusual, served three purposes. First, the double spiral halved the rotational speed needed to create a picture. This meant that the centrifugal force on the lenses in the disc dropped to a quarter of that of a single spiral. Secondly, a double spiral disc is naturally balanced – unlike a single spiral. Thirdly, the Nipkow disc could have operated as a 'closed-loop' TV system on one Nipkow disc – imaging on the downward scanning side, viewing on the upward-scanning side. Another point, though cosmetic, is that a single spiral 8-aperture Nipkow disc would give a massively arc-scanned picture with a sweep of 45 degrees. The double-spiral approach would halve that to a more reasonable 22.5 degrees.

Today we walk up to it and see this ramshackle collection of plywood, bicycle chains, 'bulls-eye' lenses, roughly cut discs distorted out of shape and a motor that looks like it came from an old washing machine. It looks just like somebody's idea of a joke. Baird achieved Television with something like this? Well, not quite. Apart from the missing photocell, this reconstructed model represented the apparatus used in April 1925 for demonstrations at Selfridge's store. It was there that Baird demonstrated the system working, using light reflected off simple shapes. The system did not however have the ability to reproduce subtle tones. That achievement came months later in October 1925 with the William Taynton story. This was but a model of an early version of equipment that led to true television.

Today though, the rather dilapidated equipment on display is only one of the two nearly complete experimental set-ups that we have from Baird. The other, more authentic but still incomplete, resides in Falkirk.[16] With little else to consider, we naturally take this equipment to be representative of what Baird was able to achieve throughout the entirety of his creative life in television – and that is our folly.

The double-8 systems in both the NMPFT and in Falkirk are good representations of what Baird did, but only in his early days. Then, he was an inventor working virtually on his own, with little funds, building equipment to demonstrate his vision – that television was feasible. His later work after he had become established is difficult to represent properly as there is so little material to go on. In a way, the double-8 apparatus is more of an embarrassment compared with the quality of his later work of the 1930s and 1940s. However, when the media want to do an item on Baird, the double-8 is about all they have to go on. As a result, when we think of Baird, we think of his creations made out of plywood, cardboard, ceiling wax and string. The 'pile of junk' label has stuck to Baird and sits uncomfortably with the knowledge that he had an important part to play in developing practical television.

This absence of any tangible hard evidence from Baird's early experimental work is tragic. An appointments diary from 1940 onwards is all that there is as a record of the events in Baird's later life.[17] Unfortunately, there are no laboratory workbooks, diagrams, charts or technical notebooks from Baird. It appears he not only did not keep them, he may not have made them in the first place. Dora Jackson, Baird's personal secretary from 1930, said in 1983, 'I'm afraid the idea of his keeping a record ... amuses me to a degree.'[18]

This means that for the experiments he completed and the demonstrations he gave almost within living memory, there is very little physical material we can use today to make an objective assessment of the contribution that Baird made to television's history. Or so it was thought until the discoveries from the Phonovision discs.

As those discoveries began to unfold, the results cried out for more background material on Phonovision. Strangely, almost no new information was forthcoming. There were still quite a few Baird people around in the early 1980s, but not one of them appeared to have been involved in making these Phonovision discs. Time had taken its toll of those earliest of the British TV pioneers.

The enthusiasts and historians before me do not appear to have emphasised the importance of Phonovision – as being the first-ever video recordings. Without images back from them, this is not at all surprising. It appeared that Phonovision had been written off. Though right throughout the 1930s, Phonovision was never declared as being a failure, the absence of any hardware or discs was very obvious. Most of the information on the process in Baird's studio in 1928 ended there with the lack of demonstrable quality. When it comes down to it, no one is interested in an idea that simply does not work.

In the early 1980s, the literature on Phonovision was scanty to say the least. The sum total was what was written in contemporary books and the press (notably the 'Television' magazine), what could be gleaned from a few official photographs, and what could be made of Baird's video recording patents. Added to that was the confusion about what actually *was* Phonovision.

The patents are probably the best place to gain an idea of the progress of Baird, or at least the progress of his ideas. For Phonovision, they are the *only* place we can go to get an understanding of how Baird's thinking on video recording developed.

Patent 289,104

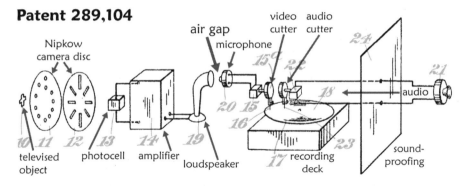

Fig 4-5. Diagram from Patent 289,104 annotated to show the recording process. The photocell and amplifier generate the video signal, which is fed to a loudspeaker. A microphone connected to the record cutter captures the video signal onto disc. A second cutter captures audio onto a separate groove of the disc.

The Phonovision Patents

Baird's first Patent on recording television was provisionally applied for in October 1926.[19] The invention called for capturing video and sound onto one conventional disc – a process that he called Phonovision in the patent. Notably, Baird does not address any special conditioning of the vision signal in this or any of his patents. In fact there is the tacit assumption that the recording and playback processes were transparent – not altering or degrading the signal in any way. Baird merely treats the vision signal from his main video output as if it were audio.

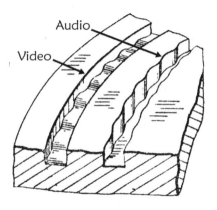

Fig 4-6. The proposed separate grooves for audio and video in Patent 289,104.

Strangely, the diagram of the connection with the recording equipment shows an air gap (see Figure 4-5). Instead of a direct connection, Baird's diagram shows a microphone in front of a loudspeaker. Even at the time this was an odd thing to do. A direct electrical link is the most assured way of getting the best quality. Some may argue that having a microphone in front of a loudspeaker could have been a means of protecting his ideas by spoofing the opposition. Whilst that is a distinct possibility, it is more likely that this arrangement reflected the real situation. Two pieces of evidence that support this come from the Phonovision discs themselves and are covered later.

Baird included two separate side-by-side grooves in the recording – one for audio and the other for the video signal (see Figure 4-6). Though the patent is largely uninspiring, it includes a suggestion of combining the separate video and audio signals into one groove. The sound would be conventionally recorded as side-to-side movement of the groove and the video would be recorded in an up-and-down movement (called 'hill-and-dale' recording).

This is essentially the principal of stereo audio recording, patented in the early 1930s by Alan Blumlein, one of the most brilliant engineering minds of the 20th century and a highly prolific inventor. Stereo recording is widely regarded today as Blumlein's greatest achievement, yet here we see that the principle of recording two channels in one groove was not new.[20] However, let us not overstate Baird's achievement. The reference to dual-track recording was made in reference not to audio, but to combined video and audio. The idea was perfectly feasible but today there is no evidence of him having tried it. Presumably, if Phonovision had been successful, he could have then explored the possibility of such a method of recording. All the early television disc recordings found so far were recorded conventionally: there were no double spiral recordings and no stereo recordings combining audio and video. The recordings are all video-only and therefore silent.

The text of the patent describes the need for coupling the record/replay deck with the scanning disc. It states:

> 'It is preferred that the driving mechanism for the record-element should be mechanically geared or otherwise so coupled to the exploring mechanism (camera) that their relative rates of move-ment can be readily reproduced. It is important for the re-production that the ratio should be exactly the same as during the recording; by mechanically gearing the record-element to the exploring device (camera) the reproducibility of the ratio of their rates of movement is made quite simple.'

In his next video recording patent, No 320,909, Baird developed the solution to synchronisation – that is, a means for maintaining a stable image display – by directly linking the Nipkow scanning disc to the record platter through a system of gears (see Figure 4-7).[21] So long as the disc was played back on the same arrangement used for recording, the replayed picture would be stable, or ought to have been.

For every turn of the Nipkow scanning disc, the turntable would turn through a precise fixed angle. That meant that the video recorded on the disc would have a direct angular relationship with the respective aperture on the Nipkow disc. In this way, the position of a certain point in the TV frame would relate to a certain angular position around the disc recording. By doing this, Baird was hoping to avoid the difficulty of dealing with timing information either as a separate additional channel or, like today, embedded within the video signal. In either of these cases, reliable timing extraction from a signal recorded on a disc in the late 1920s in the 'toddlerhood' of electronics would have been far too great a challenge.

The diagram in the patent is made complicated by Baird including a schematic of clutches and gearing that was going to be needed to make the system work. Baird was demonstrating that he understood not just synchronisation but also the need to gear down the fast-turning Nipkow disc to a rate that the record platter could handle. The reduction ratio measured directly from the diagram is 10:1 (a first stage of 5:1 then a second stage of 2:1). That is, ten turns of the scanning disc (10 TV frames), would be recorded on every turn of the record. With Baird's standard dictating a Nipkow disc turning at 750 rpm (12½ times per second), the gearing would turn the record platter at 75 rpm – close to the 'standard' recording disc speed of 78 rpm. Whether Baird had intended to provide so 'correct' a diagram in his patent is not clear. However, it does show that he

Patent 320,909

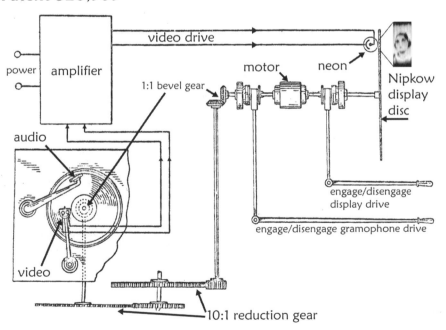

Fig 4-7. Patent 320,909 expands on the suggestion in Baird's earlier patent for synchronisation of scanning disc with record deck.

was giving serious thought to dealing with synchronisation, despite the solution being complex.

There is one feature of his patent that raises an eyebrow. Baird included clutch mechanisms in his drive. He declared in the patent that this would allow independent operation of the scanning disc and the record platter. The whole point about Phonovision was for it to operate in perfect synchronisation. If this was the intention with this patent, then it should not require any clutch mechanisms. Indeed, they may even have degraded the performance. Baird probably incorporated clutches to allow the picture to be correctly framed.

Patent 324,049

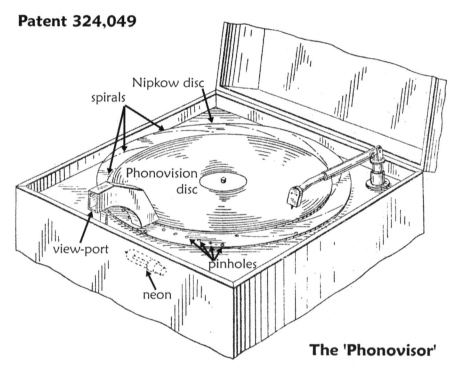

Fig 4-8. The innovation of the 'Phonovisor' is in using a Nipkow display disc for the gramophone deck. A conventional playback speed of 78 rpm would have dictated the use of multiple spirals as shown here. So long as the video recording was captured on a similar system, the playback of the video would be stable, not requiring any other means of synchronisation.

The Phonovisor Patent

By the time of his next video recording patent, there appears to have been a radical change in Baird's thinking. Gone are all the gears.[22] We are left with a playback device, which Baird called a *Phonovisor* for viewing

Phonovision discs, not much more complicated mechanically than a standard gramophone player. As there were a fixed number of 30-line frames on each revolution of the Phonovision disc, each TV line occupied a unique angle around the disc. His synchronisation was embedded in the recording.

Instead of having a geared linkage between the record platter and the Nipkow scanning disc, Baird took the mental leap of having the Nipkow disc mounted co-axially underneath the record platter (see Figure 4-8). It had a larger diameter than the platter, with the spirals of apertures of the Nipkow disc appearing outside the edge of the record platter. The special Nipkow disc had exactly the same pattern and number of apertures as there were frames recorded on the disc. The pick-up

> ## Phonovisor Practicalities
>
> Videodiscs made to run at 75 rpm would have required ten 'spirals' of apertures around the periphery of the Nipkow disc. Such a large number of spirals is suggested by the illustration on the 'Phonovisor' Patent where we can count ten of them. For a 30-line picture, ten spirals would mean 300 apertures. For, let's say a 38 cm (15 inch) diameter Nipkow disc, this would mean that the physical length of a TV line would be around 4 millimetres. To see sharp detail, each aperture would have to be precision drilled and positioned with great accuracy – probably to around 100 microns. The patent of course describes the principal of the invention – not necessarily the hardware implementation.

for the video signal would be amplified and fed to a neon lamp under the Nipkow disc, shining through the apertures. Looking down on the disc, we would have seen the tiny picture form as the disc turned. With correct alignment of the disc on the turntable, the picture would have been perfectly stable, independent of changes in the motor speed.

His next patent on video recording appears today to be disappointing and almost too simple to be worthy of being considered an invention. British Patent No. 320,687 describes nothing more than a changeover switch for a combination gramophone and television receiver.[23] One position of the switch let video and audio through from the television receiving section, the other position let audio through from the gramophone section. In engineering terms, this was nothing more than a dual-pole changeover switch constrained by the patent to video and audio.

The Thinking behind Phonovision and the Phonovisor

In these few patents spanning 1926 to late 1928, we can see the basics behind Phonovision: a process for recording using a camera and recording system mechanically linked and synchronised and a mechanism for playback – the Phonovisor – brilliant in its simplicity.

If this could be made to work, Baird would have the capability to deliver not just inexpensive playback units but the recordings to go with it.

Though Baird has said himself that he was not a businessman, the solution he thought up for the playback unit would have been an effective mass-market device (see Figure 4-9). If it had succeeded, the Phonovisor would have been the lowest cost videodisc player with integral display in the history of television (see Figures 4-11 and 1-1).

Publications

The first issue of the monthly journal, 'Television', appeared in March 1928. It is today a sought-after source of reliable contemporary information on engineering developments in television, despite its rather obvious bias towards to Baird. In the June 1928 issue, Baird reported on Phonovision:

> '... we have now succeeded in getting ... images *(back from the disc)* and we can see ... a crude smudgy replica of the person whose image has been put on the gramophone record ... at the present time it is more of a curiosity...',[24]

Presumably, this was more a curiosity than a practicality. Alfred Dinsdale, author of the earliest book on television (1926) and the first editor of 'Television' magazine, reported in 1928 that, 'Baird ... has initiated many experiments in the recording of image sounds...' (see Figure

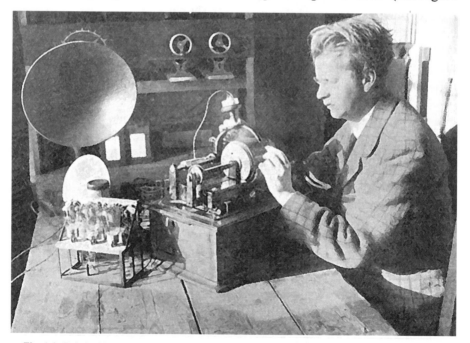

Fig 4-9. Baird with phonograph equipment. The original caption by Barton Chapple refers to Baird 'carrying out his original Phonovision recording experiments'.
From 'Popular Television', Barton Chapple, 1935

4-10) and that Phonovision was, 'still in the laboratory stage of its existence'.[25]

In a review of Baird's achievements at the end of 1929, Dr Clarence Tierney described Phonovision.

> 'Phonovision is the method evolved by Baird by which the image sound is permanently recorded. By means of these records the original scene can be reproduced repeatedly in the Televisor at any time, and the significance and value of this method for the storage and reproduction of the original images of living and other subjects are obvious and as important for vision as is the gramophone record for speech and music.'[26]

If accurate, this report suggests that several video recording experiments were operating in parallel. We would certainly expect at least two – for exploring both the recording and playback processes (see Figure 4-11).

By 1931, little appeared to have changed. Moseley and Barton Chapple reported in their book:

Fig 4-10. Detail from a photograph of Baird's apparatus at the British Association meeting in Leeds in 1927 showing Noctovision and cylinder Phonovision equipment.
Courtesy of the Royal Television Society RTS 36-34

'… at present, (Phonovision) is merely a scientific curiosity…'
and that, '… the Baird Company are pursuing their investigations
with Phonovision so that the apparatus might be perfected.'[27]

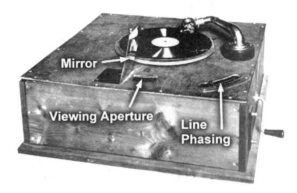

Fig 4-11. A mock-up of the Phonovisor apparatus.

Courtesy of the Author

Phonovision and the Phonovisor largely dropped out of news reports
from then on. In 1934, there was a hint of a development period around
1928.

'The principles and practice of (Phonovision) were established
about six years ago…'[28]

A search through all the available literature resulted in several mentions
of Phonovision but no mention of demonstrations of playback of pictures
from the discs.

In all the literature there are just a handful of photographs that are in
some way related to Baird's television recording experiments. There are
three separate views of what is called the Phonovision studio (Figures 6-3
to 6-5), one of apparatus with W. C. Fox and J. D. Percy (Figure 6-10), one
of a Phonovision disc (Figure 6-2) and three or four of a mock-up of the
Phonovisor playback unit (Figures 1-1, 4-11). The pictures all appeared in
1928 or earlier. There were no later pictures. There is an odd photograph of
Baird sitting by a cylinder record/playback unit (Figure 4-9) and one with
him at a meeting of the British Association at Leeds in 1927, also with a
cylinder record and playback device (Figure 4-10).

Personnae

The contemporary literature, the timing when the photographs first
appeared and the dates on the discs all suggest a period of between 1926
and 1928 for Phonovision. With such a period almost in living memory, it
became vital to find and talk with the early pioneers. The quality of

information from the few surviving individuals illustrated that reliability of memory after such a period, without the benefit of something like a diary, is fairly low. Some were claiming events that could not possibly have happened at the time they said. Rather than anything sinister, this is just the failing of memory from which we all suffer.

Witness: B. Clapp

The oldest surviving members of Baird's team were W.C. (Bill) Fox and Ben Clapp (see Figure 4-12). I first met Clapp in 1982, as he was the owner of one of the Phonovision discs. In 1926 Clapp and a business colleague had approached Baird with a proposal for a demonstration of television transmitted across the Atlantic.[29] After Clapp had finished with his wireless business, Baird hired him and set him to work on demonstrations of television reception over long distances.

Though he owned a Phonovision disc, Clapp had no knowledge and claimed to have had no involvement in video recording. When Clapp joined the company in November 1926, Baird had two other assistants – J. J. Denton, a Physics lecturer from Morley College whom Baird had met earlier in Hastings, and the 'invaluable' Wally Fowlkes. Clapp recalled that there had been some problem with Phonovision but could not elaborate. At the time, he said, he was heavily involved in long-distance test transmissions.

Fig 4-12. Ben Clapp in April 1924 prior to joining Baird and when he was working with Wanamaker's.
From original courtesy of R. M. Herbert

In late 1927 and early 1928, Clapp had been working for Baird in the USA, in New York State. He was in the company of a radio amateur, Robert Hart W2CVJ, setting up and testing the receiving end of a possible demonstration of receiving television images from his own transmitter in Coulsdon, Southern England.

After Clapp died, his life-long friend and ex-Baird employee, Ray Herbert, found the log book for G2KZ, Clapp's amateur radio station. Herbert discovered that a few entries referred to disc recordings. One entry (24th November 1927) refers tantalizingly to 'face and hand', echoing the content of the earliest-known Phonovision disc.

An earlier entry in Clapp's logbook, made shortly after midnight on 7th October 1927, gives the message from Baird, 'Pse (please) stand by for TV record.'[30] (This is probably the first-ever mention of the abbreviation 'TV' for television).[31] Later that same day, Baird had given a lecture at the White Rock Pavilion in Hastings in which he demonstrated playback of videodiscs.

> 'The audience then had the novel experience of listening to gramophone records of the sound made by a dummy's head as produced by Mr Baird's apparatus. He went on to explain how the sounds were reconverted into a picture by an apparatus which is practically the reverse of the transmitting apparatus.'[32]

Clapp's Phonovision disc was date-stamped 20th September 1927, and we now know that it contains images of a ventriloquist's dummy head. The transatlantic transmission early on that Friday, 7th October was the first ever entry in the logbook for a transmission of a 'TV record'. It is just possible that Baird took the disc to the Hastings lecture later that same day. Stretching conjecture a little further, Clapp's Phonovision records may well have been mementoes from their use in test transmissions across the Atlantic.

Witness: W. C. Fox

Bill Fox (see Figure 4-13) had been an Associated Press journalist and a supporter of Baird since the early days. He was present at the first demonstration of television on 26th January 1926 to members of the Royal Institution and acted as a 'minder' ensuring that no more than six at a time entered the small demonstration room in Baird's premises in Frith Street.

His association with Phonovision comes from a photograph of him, posing with J. D. Percy and some lab equipment. Unfortunately in 1983, he had no recollection of the event. The only memory he had was of a problem during recording.

Fig 4-13. Bill Fox being televised during the Transatlantic transmissions in 1927–28 at Long Acre.
Courtesy of the Royal Television Society
RTS 36-25

However, he advised me that his lack of technical understanding could give rise to inaccuracies in his recollection.

> 'Around that time (1927 or 28), I remember wandering in to a laboratory one late afternoon and watching a run of television

images from disc recordings. Towards the conclusion of it, one of the engineers present remarked "That seemed a very short display to me. Was it cut for any reason?" The remark came from one of Columbia Graphophone Co.'s engineers, but on investigation it proved to be a standard recording of normal length and the "shortness" of the display was traced to the disc slipping on the turntable due to there being no felt covering on the turntable's polished steel surface!'[33]

Notably, Bill Fox recalled engineers from the Columbia Graphophone Company being present. This is the name sported on the Phonovision Test Record labels. If indeed there were discs slipping on the smooth turntable platter, none show that feature today.

By 1982, the key individuals who would have had direct knowledge of Phonovision – J. D. Percy and Harry J. Barton Chapple – had long gone. The other early Baird engineers had little knowledge or had started late.

Witness: J. D. Percy

James D. Percy (see Figure 3-18) joined the Baird Company along with many of his colleagues in early summer of 1928. He was involved in demonstrations both at home and abroad and in the outside broadcasts of the Derby from the Epsom racecourse. He also was one of the few Baird engineers that had an involvement with Phonovision. Fortunately, unlike his contemporaries, he recorded a few observations on Phonovision in his unpublished memoirs, referring to it as one of Baird's failures.

'Phonovision, the undoubted father of all video recording systems, simply meant making a gramophone record of a television signal instead of music, and playing it back through a Televisor instead of through a loudspeaker. The results, from a standard wax disc running at 78 rpm, were of course appalling and only at brief intervals, and by using a lot of imagination could anything approximating to a recognisable image ever be identified. However, this was enough for Baird and his publicity seekers. "Recorded television is here" screamed the headlines, "In a darkened room in Long Acre I have seen the first glimmerings etc. etc..."'[34]

Percy's observations on the quality of Phonovision are accurate and describe exactly what we see today. He stayed with the Baird Company until 1939.

Witness: T. H. Bridgewater

Thornton H. (Tony) Bridgewater (see Figure 4-14) had joined the Baird Company in 1928 and had been sent out to Australia for six months to promote Television there. On his return, he worked on a variety of projects, but none involving video recording. He was however involved in the start of regular television broadcasting in 1929 and in large screen displays in 1930. He told me that as far as he knew the work on Phonovision had finished by the time he returned from Australia in early 1929.

This is in agreement with the absence of further patents and contemporary literature. Bridgewater mentioned though that he knew of someone else who claimed to have been directly involved with Phonovision. His name was C. L. Richards.

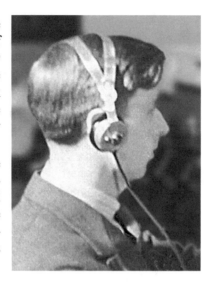

'He was in the company before me, and when I referred to Phonovision in a lecture some years ago he was in the audience and got up afterwards to say he had been present during the recordings.'[35]

Fig 4-14. 'Tony' Bridgewater at the sound control desk in BBC Portland Place on 4th April 1935.
Courtesy Royal Television Society
RTS 37-84

Witness: C. L. Richards

'Dick' Richards (see Figure 4-15) had indeed been working on some form of video recording in late 1928 but work on 'Phonovision' had already finished.

> 'I joined Baird in Autumn of 1928, and having some knowledge of electrical recording, spent a short time on "Phonovision" *(Richards, like many others, refers to any generic recording of television in Baird's time as 'Phonovision')*. The Crystalate Gramophone Record Co. helped me with material, but I do not recall the recording machine itself being there or whether I made it, neither the speed of my 12 inch blanks.'[36]

> 'My recordings were definitely made on wax, and would have replayed from wax. These wax blanks were 1½ inches thick and 12½ to 13 inches diameter … I also used the "conventional synchronising pulse" *(and goes on to specify the 3% black level period of Baird's 30-line standard)* and no mechanical linkage between scanning disc and record.'[37]

> 'Am fairly certain no shellac pressings were made, this being a very expensive process. The cutter was probably mine, and also I expect the replayer, straight from the wax, once only. The pictures were not very good, my problems being mechanical resonances…'

Richards' recordings were standard 78 rpm onto soft wax without any mechanical linkage between turntable and scanner. These therefore do not conform to the definition of Phonovision. Nevertheless, this story shows that Baird was interested in seeing what an expert like Richards could do.

Referring to the Phonovision discs, he continued;

> 'So your 10-inch … recordings pre-date my arrival at Long Acre, and I cannot recall their existence. That they have been identified as Columbia suggests to me that Columbia may have done the recording. Although I know the Columbia recording engineers personally, they never mentioned it.'

Richards' lack of knowledge about the Phonovision discs and the contemporary photographs of the equipment labelled as being for Phonovision led him to say,

> 'All this strongly suggests to me, that the recordings were made several years before my time…'

This lack of knowledge is puzzling. We would expect that someone coming into the organisation with a specialisation in electrical audio recording would have been at least briefed on a previous attempt. The explanation that fits best is that Richards was simply not told of the previous Phonovision experiments and that Baird was interested to see if Richards could come up with anything different. He spent only a very short time on this before he was moved on to other things – notably building equipment for transmitting 30-line television from Witzleben, Germany. By 1932, he had left the company.

Fig 4-15. 'Dick' Richards in Berlin 1929 as part of the Baird engineering team.
Courtesy of R. M. Herbert

Witness: F. Whitworth

One other person claimed to have direct knowledge of Phonovision. This was Frank Whitworth who had approached Granada TV during their research for the 13-part series entitled 'Television', broadcast in 1985.

In an exchange of letters, Whitworth recollected,

> 'I was assistant to Mr Barton Chapple, about 1931, when he was engaged with "Phonovision", but not much success, as many difficulties were experienced with the high reduction gearing from the disc to the turntable, but some discs were made, Sir Oliver Lodge was very keen with John Logie to go ahead with "Phonovision", and visited the lab on many occasions. The

turntable speed was approx 16 to 25 rpm to the best of my memory, I was 16 years old at the time.'[38]

The date seems rather late for Phonovision but does tend to agree closely with Barton Chapple's report in his 1931 book that work was continuing. It might be that Barton Chapple was making another attempt to playback pictures from the already-recorded Phonovision discs. Unfortunately, there is no more evidence.

Bridgewater verified Frank Whitworth's work at the Baird Company and that he had been, 'quite a youngster at Baird's and helped in the stores'.[39]

J. Gilbert

The DRS (Developments in Recorded Sound) archive holds a taped interview with John Gilbert made in 1984.[40] In it, Gilbert said he made probably the first ever recording of a television programme in 1927 using a Baird *Televisor* that he borrowed from Baird. Gilbert went on to say that he returned the equipment and told Baird that he had made a recording. He also said that a few days later the Baird Company took out a patent on recorded vision, though it had been his idea rather than Baird's.

This is a surprising claim, and distinctly puzzling. In 1927, there were no television programmes. The first ever programme was broadcast by Baird in July 1928 from his offices at Long Acre. The first television transmission through the BBC's 2LO transmitter took place on one occasion only in March 1929. Broadcasting started in earnest in September 1929 with the Baird service through 2LO.

Borrowing a Televisor from Baird would have been difficult and unlikely – not just because of his careful attention to commercial security. In 1927, Baird had no more than a 'fretwork' prototype Televisor with the commercial Televisors some three years away. In 1928, Baird built around twenty 'Noah's Ark' Televisors. More than ten of these went to the company directors and the rest were used for demonstrations. These Televisors were much larger than the later commercial offerings, using 24 inch (61 cm) diameter Nipkow display discs. Finally, the provisional patent on recorded vision was dated October 1926. From the best information available, Baird had moved on from Phonovision by 1929.

A Wellsian Future from the 19th Century

In the late 19th and early 20th centuries, arguably the most accurate visionary of the future was Herbert George Wells. His 1899 novel, 'When the Sleeper Awakes,' is set in London in 2099 and describes a small elite core of society unilaterally imposing their version of Utopia on all. The story has a wealth of predictions. In it, he foretold of air conditioning, automatically activated doors, war in the air between aircraft fitted with guns and bombing from aircraft. He also presented the idea of portable television sets, video recordings and video libraries. In this story, books no longer exist, having been replaced by audiovisual recordings.

> 'The hero observed one entire side of the outer room was set with rows of peculiar double cylinders inscribed with green lettering on white that harmonised with the decorative scheme of the room, and in the centre of this side projected a little apparatus about a yard square and having a smooth white face to the room. A chair faced this. He had a transitory idea that these cylinders might be books, or a modern substitute for books, but at first it did not seem so.'

> 'He puzzled over a peculiar cylinder for some time and replaced it. Then he turned to the square apparatus and examined that. He opened a sort of lid and found one of the double cylinders within, and on the upper edge a little stud like the stud of an electric bell. He pressed this and a rapid clicking began and ceased. He became aware of voices and music, and noticed a play of colour on the smooth front face. He suddenly realised what this might be and stepped back to regard it. On the flat surface was now a little picture very vividly coloured, and in this picture were figures that moved. Not only did they move, but they were conversing in clear small voices. It was exactly like reality viewed through an opera-glass and heard through a long tube.'[41]

The double cylinder was no doubt inspired by the development of the audio cylinder, probably 'double' to carry vision and sound. Apart from it being uncannily like current pre-recorded video discs and tapes, it is also a good prediction of the work that John Logie Baird undertook, some 25 years after the book was published, in developing a video recording system. Before Baird started developing Phonovision on discs, he experimented first with recording video on cylinders. This of course was an inevitable outcome of the availability of recording equipment in 1926, when the idea came to him.

Could it be that Wells' story was influential on Baird, inspiring him to try out video recording? As far as we can tell, no other television pioneer, not even in the United States of America, had tried to make such video recordings in the late 1920s. If they did, they neither advertised the fact nor kept any record of such experiments.

Baird worshipped Wells as his hero. His tales of a visionary future filled with new and exciting technologies in a Utopian society grabbed Baird's imagination. He would no doubt have read of many of the fanciful inventions and ideas liberally spread throughout the books, including that of the 'sleeper's video recordings and video library.

Professor Malcolm Baird reflected on his father's inspiration. 'What we think is that his interest started with reading the scientific short stories of H. G. Wells and I think he devoured these books and was inspired by them.'[42]

In September 1931, Baird sailed to the United States on board the 'Aquitainia' and discovered that his hero was on the same ship. 'Among the passengers was H. G. Wells and I was quite excited at the prospect of meeting a man who in my youth I had regarded as a demigod.'[43] His companion, Mr Knight, arranged a meeting on deck – a meeting that Baird as a youth could never have imagined he would make (see Figure 4-16). He however found Wells quite ordinary, describing him as, 'a poor vulgar creature like myself. We had a short chat about youth camps. I said these organisations appear to ignore sex. "Oh well" he said, "Every Jack has his Jill," and that is all I remember of the conversation with my demigod.'[44]

An article in the 'Television' journal of 1929 drew attention to the similarity of Wells' vision and Phonovision, although Wells' influence on Baird was not suggested. The closing remarks of the author are strangely prophetic.

'The printed word will still be necessary for many things, but thrillers, shockers (horror), and other inhabitants of the libraries

Fig 4-16. H. G. Wells (left) and J. L. Baird (right) meeting on the deck of the Aquitainia, September 1931.

Courtesy of R. M. Herbert

will be scenarioed by the author and phonovised by the publisher. So every house may be the home of a theatre.'[45]

Fictional Phonovision

The ability to record and playback television caught the public's imagination. In July 1929, Derek Ironside wrote a short detective story where the criminal was caught through the use of a Phonovision video recorder. When presenting the evidence captured on the video recorder in court, the apparatus is described.

> 'Our television apparatus can be adjusted to take a permanent record of anything it receives. The electrical impulses constituting the pictures received are converted into sound waves and recorded upon a sort of gramophone-disc, this process being well known to experts as phono-visual reception. To reproduce the picture, a gramophone pick-up is used in conjunction with a television receiver.'[46]

Using image detail and slow-motion replay reminiscent of today's video recorders (and both quite impossible to achieve with Phonovision), the video recorder provided the evidence to identify the criminal. Most of these short stories by Ironside hinged on Baird's latest developments, and were uncannily accurate on their eventual use years later. Baird's developments were simply not good enough for the applications in the storylines. It took decades for video recording, infrared surveillance television and colour television to get anywhere near the practical applications in those early fanciful stories.

Experiments and Demonstrations

Phonovision was just one of a series of developments as part of Baird's exploration of the possibilities of this new medium. This exploration took place roughly in the period between 1926 and 1929. Each of the results of his work culminated in a demonstration with no apparent follow-through. In that sense, they were true experiments, with no regard for practical implementation.

Recorded television made its debut on 6th January 1927. Baird was presenting a lecture to the Physical and Optical Society of Imperial College, London.[47] He illustrated his lecture with a playback of a recording he had made of his video signal. He described how the quality of the sound changed depending on the object being televised.[48] In the following months, Baird continued to present demonstrations of the playback of the sound of recorded television. The public lecture at the White Rock Pavilion in Hastings during the 'Wonders of Science' exhibition in October 1927, cited earlier, is one other such occasion.

'Phonovid'

In 1965, a technique for storing scanned pictures onto gramophone records was developed by Westinghouse Research Labs in Pittsburgh, Pennsylvania. Using a conventional vinyl LP disc, the approach stored around 400 monochrome still pictures, each of 360 lines and taking 6 seconds to replay each picture. The image appeared on a standard television monitor, the picture 'held' in a special scan converter. The system was claimed to be for television pictures.

Like Slow Scan TV, this system was only designed for isolated stills, captured from a television camera. It was not in itself handling Television.

Farr, K E, 'Phonovid – a System for Recording Television Pictures on Phonograph Records', Journal of the Audio Engineering Society, 16, No 2, April 1968, pp163–167

New York Times, 'Westinghouse Putting TV on Phonograph Records', 6th May 1965

Whilst there are plenty of reports of the audio playback of recorded vision signal, nowhere is there a report of a demonstration of video playback. Was there a problem?

In his autobiography, Baird talks about Phonovision and the disappointment it brought him.

> 'I got a gramophone record made of these sounds and found that by playing this with an electrical pick-up, and feeding the signal back to a television receiver I could reproduce the original scene. A number of these records were made (and one can be seen at the Science Museum in South Kensington) but the quality was so poor that there seemed no hope of ever competing with the cinematograph.'[49]

The contradiction in Baird's words – 'I could reproduce the original scene' and, 'the quality was so poor' – is worthy of note. His claimed ability to reproduce the original scene is challenged by the failure of anyone else to do so from those very Phonovision discs and by the poor quality of the recorded signal. Replayed on pre-computer age equipment, Phonovision would have been simply unwatchable, offering only the occasional glimpse of something recognisable. In that sense, the observations in J. D. Percy's memoirs, cited earlier, are far more accurate than Baird's own recollections.

Baird admits Phonovision was a failure – but it was a disappointing failure in its own right, and not just because it had to compete with the cinema. After all, even *broadcast standard* 30-line television had no hope of competing with the cinema.

Fig 4-17. The Baird Company's Kingsbury Manor facility for Televisor overhaul and stores. Harry Barton Chapple (standing) points into the open back of a 'Noah's Ark' Televisor. (April 1929). Through his books and articles, Barton Chapple did much to promote gramophone recordings of the 30-line vision signal.

From original courtesy of R. M. Herbert

The Sound of Vision

Despite the lack of success with Phonovision, Baird demonstrated playback of the recorded Phonovision video signal as a sound. But why would Baird do this? At first, this appears quite bizarre. This odd type of demonstration seems in some way related to the apparently even stranger behaviour of Baird as described in his autobiography,

> 'In testing out the amplifiers I used to use headphones and listened to the noise the vision signal made. I became very expert in this and could even tell roughly what was being televised by the sound it made. I knew, for example, whether it was the dummy's head or a human face. I could tell when the person moved, I could distinguish a hand from a pair of scissors or a matchbox, and even when two or three people had different appearances I could tell one from the other by the sound of their faces.'[50]

For the makers of sound-bite documentaries focusing on the obscure and the nutty, this is indeed great material. The bizarre-ness however is false. After spending years processing the 'sound' from Phonovision discs and the later 30-line broadcast material, I can say that you *do* get into the way of recognising and discriminating certain actions. You really can tell whether you have a live subject (the sound changes all the time and changes smoothly) and a static subject (no change). For dancers, you can even tell the tempo as the sound pattern becomes periodic if the dancers movements are repetitive. Baird was not making this out to be something special – it was just a curiosity and certainly not intended to be a valid way of 'watching' television as could be inferred.

Baird demonstrated the sound from Phonovision more than likely because he was unable to show pictures. That Baird still demonstrated sound playback from Phonovision may well be because of two reasons. First, I believe Baird wanted to suggest that work was progressing, thus boosting confidence in his ability to deliver television as a service (which, prior to 1929, he had yet to do). Second, the demonstration would have been useful in priming potential purchasers about what they should expect to hear on their radios when the broadcasts eventually started (see Figure 4-17).

Despite sporadic tantalising comments throughout the early 1930s, the promise of Phonovision fizzled out. As time passed, the test discs were largely forgotten for what they represented and became curiosities. The Phonovisor – the world's first videodisc player – became an unfulfilled dream.

---

[1]Memo from 'Ian M.' to 'R.L.B' dated 14th Apr 1981, sourced from T. H. Bridgewater, 1982

[2] PARR, G.: Private correspondence, 4th Feb 1960

[3] PARR, G.: Private correspondence, 15th Feb 1960

[4] 'We seem to have lost the Picture' (BBC LP from the series '40 Years of Television'), 1976, REB 239

[5] LEGGATT, D. P.: Private correspondence to Author, 12th Oct 1983

[6] GEDDES, K.: Science Museum, Private communication, 1st Aug 1982

[7] BRIDGEWATER, T. H.: Private correspondence to Author, 7th Dec 1983

[8] LEGGATT, D. P.: Private correspondence to Author, 8th Dec 1983

[9] ANON.: 'The Birth of the Television Society', *Television*, Vol 1, No 1, Mar 1928, p16

[10] Science Museum Inventory 1935-335

[11] MCLEAN, D. F.: 'Computer-based analysis and restoration of Baird 30-line television recordings', *Journal of the Royal Television Society*, **22/2**, Apr 1985, pp87–94

[12] BRIDGEWATER, T. H.: Private correspondence to Author, 9th Mar 1985

[13] '70-year-old video gets its first play', *The Times*, 28th Dec 1996

[14] 'After 70 years, playback for Baird's video recordings', *Daily Mail*, 30th Dec 1996

[15] FOX, W. C.: Interview extract, 'Television' (Granada TV), 1985, episode 2

[16] WADDELL, P.: 'John Logie Baird and the Falkirk Transmitter', *Wireless World*, Jan 1976, pp43–46

[17] BAIRD, M. H. I.: Private correspondence to Author, Aug 1999

[18] JACKSON, D. K.: Private correspondence to Author, 14th Sep 1983

[19] BAIRD, J.L.: British Patent 289,104, applied for 15th Oct 1926

[20] BLUMLEIN, A.D.: British Patent 394,325, applied for 14th Dec 1931

[21] BAIRD, J.L.: British Patent 320,909, provisional, complete specification

[22] BAIRD, J.L.: British Patent 324,049, applied for 10th Oct 1928

[23] BAIRD, J.L.: British Patent 320,687, provisional, complete specification

[24] BAIRD, J. L.: 'Report of the 1st General Meeting', *Television*, June 1928

[25] DINSDALE, A.: 'Television' (Television Press Ltd), 2nd edn, 1928

[26] TIERNEY, C.: 'The Origin and Progress of Television Part II', *Television*, Jan 1930, p546

[27] MOSELEY & BARTON CHAPPLE, 'Television – Today & Tomorrow' (Pitman), 2nd edn, 1931

[28] BARTON CHAPPLE.: 'Canned Television', *Practical Television*, 3rd Mar 1934 (pre-release article)

[29] BAIRD, J. L.: 'Sermons, Soap and Television' (Royal Television Society), 1988, p88

[30] CLAPP, B.: G2KZ Logbook entry for 7th Oct 1927, reproduced by Ray Herbert, 1998

[31] HERBERT, R. M.: Private communication to Author, Feb 2000

[32] *Hastings & St Leonards Observer*, 8th Oct 1927, information supplied by R. M. Herbert, June 1994

[33] FOX, W. C.: Private correspondence to Author, July 1983

[34] PERCY, J. D.: 'The Vision Machine' (Unpublished memoirs), Aug 1979, p55

[35] BRIDGEWATER, T. H.: Private correspondence to Author, 3rd Feb 1984

[36] RICHARDS, C. L.: Private correspondence to Author, 12th Jan 1984

[37] RICHARDS, C. L.: Private correspondence to Author, 12th Feb 1984

[38] WHITWORTH, F.: Private correspondence to Author, 1985

[39] BRIDGEWATER, T. H.: Private correspondence to Author, 20th Jan 1985

[40] British Library, National Sound Archives, DRS 26, C90/54/01, Interview conducted in May 1984, information supplied by Prof Robert Maconie, Savannah College of Art and Design, 3rd Sep 1998

[41] WELLS, H. G.: 'When the Sleeper Awakes', 1899, quoted in *Television*, July 1929, p246

[42] BAIRD, M. H. I.: Extract from interview for documentary 'Seeing by Wireless', transmitted on BBC Radio 2, 21st Oct 1997, producer Kenris MacLeod

[43] BAIRD, J. L.: 'Sermons, Soap and Television' (Royal Television Society), 1988, p122

[44] BAIRD, J. L.: *ibid,* p122

[45] ADCOCK, E. P.: 'Phonovision Foretold', *Television*, July 1929, p246

[46] IRONSIDE, D.: 'How the French Police Proved Phonovision', *Television*, July 1929, pp266–269

[47] 'Images recorded as sounds, a phenomenon of television', *The Times*, 7th Jan 1927

[48] 'Faces and Noises affect Television', *New York Times*, **II**, 23, 9th Jan 1927

[49] BAIRD, J. L.: 'Sermons, Soap and Television' (Royal Television Society), 1988, p63

[50] BAIRD, J. L.: *ibid,* p63

5   Restoring Vision

'Each lonely scene shall thee restore;
For thee the tear be duly shed,
Beloved till life can charm no more,
And mourned till Pity's Self be dead.'

W. Collins c.1750

Climbing Mountains

If it were not for computer technology, Baird's *gramophone videodiscs* would continue to be curiosities that merely hinted of a time before television as we know it. Their latent images would remain unseen and the information embedded in them would still be completely unknown.

The success of the restoration and the wealth of information that has arisen from it make this one of the more surprising and unusual finds in the history of technology. Using the right tools for the right job at the right time was the key to this achievement. The job here is complex and singularly unique: it entails studying several different format discs originally designed for audio, analysing recorded signals, understanding what the signal contains, identifying the defects, separating out the different types of defects, tackling each type and correcting for it. Once that is complete, the corrected video data needs to be re-formatted for display on standard graphics display equipment.

Line by line, the correction values plot out the profile of errors in the signal's timing. This list of corrections applied is analysed to gain clues as to what might have caused the problem that we have rectified. The job then is a series of connected and dependent tasks, all of which are completely new.

In attempting to cover all those techniques in the space of one chapter, there is a risk of being superficial. However, it is worth doing so, if only to remove some of the *mystique* surrounding the restoration. From just a few old shellac and aluminium discs to what we understand today is a remarkable leap. Gaining an idea for the techniques that were used show that this leap is just a series of steps. As an analogy, the way to climb a mountain is to put one foot in front of the other. On a mountain walk, we follow the track beaten down by the feet of the folk before us. However, for restoring these ancient recordings there was no trodden path. The process though is the same: it just takes a little longer. So let us take a journey through all the main stages of restoration. The details of the processing have been omitted, as they would occupy a complete book on their own. For a more in-depth understanding, the reader is advised to study the many current books on the subject, as listed in the Bibliography.

Fig 5-1. A Phonovision disc photographed in 1928, displaying radial spokes and carrying a 'Columbia Graphophone Company' label.
Courtesy Royal Television Society RTS 36-23

The video material is embedded in a long obsolete television format recorded onto a medium for which it was not designed (see Figure 5-1). For each problem we come across, the tools to tackle it need to be developed, built and tried out. There are no easy answers to tackling this unusual material and no ready-built, 'off-the-shelf' solutions.

Before touching on how the problems were handled, we need to gain an understanding of what the task entails. That means understanding what the television signal is and how the television image is formed. We can then start to understand what the vision signal used to be like and what caused the signal on the discs to be so degraded. This will guide us to choosing the tools we need to tackle those degradations, with the aim of restoring the television images to as close to the original quality that we can get.

Vision by Electricity

The more that television technology has matured, the more distant has our understanding of television become. That is true for most technologies today: simple concepts wrapped up in a complex solution. When the trappings of the technology solution and the 'techno-babble' that enshrouds it are removed, we see the basis for practical television systems revealed in their simplicity.

We came across this earlier, when we saw that television comprises just a few functional elements: a camera to convert the scene to an electrical signal, and some means of transmitting the signal, receiving it and converting it back to an image on a display. The core element in any television system is the video signal. That is, if you like, the image in its electrical form. This one signal however contains all the information necessary to re-construct the image. How can this be?

There is one example from the early days of facsimile that illustrates how an electrical signal can carry an image. At that time, pictures were sent between facsimile or fax machines very successfully, in what amounted to a slowed-down version of television. Only one picture was sent at a time, taking several minutes.

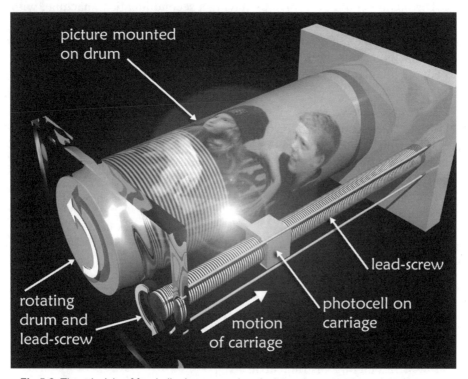

Fig 5-2. The principle of facsimile drum-scanning. A picture is wrapped around a drum. The drum is spun and a photocell, mounted on a lead-screw, slowly traverses the length of the drum. The same mechanical arrangement is used for receiving and recording the transmitted picture.

Courtesy of the Author

The picture was wrapped around a long cylinder or drum that was wide enough to carry the biggest picture to be scanned (see Figure 5-2). The drum was motor-driven and was spun on its axis quite fast, sometimes at a few turns per second. The picture appeared as a blur. By looking at just one part of the picture spinning by, as if looking down a straw, all that could be seen was a spot flickering in brightness as the picture turned on the drum. If instead a photo-detector were 'looking' at that spot, the signal from it would vary rapidly with the brightness. A detector fixed in position would produce a signal that repeated every turn of the cylinder, interrupted by the brightness step from the edge of the picture. This signal over one turn of the cylinder would be like looking along one thin swath or *line* of the image. This would be converted by the photo-detector into a video waveform that would repeat on every turn of the drum. By slowly moving the photo-detector on a carriage down the length of the cylinder, it would have traversed or 'scanned' the picture into a succession of lines.

The picture could be copied by having a second identical machine, with recording paper wrapped around the drum. In place of the photo-detector would be some method of marking the paper. The stronger the signal, the darker or lighter the mark made. However, to get the original and the copy to appear identical, both machines would need to run in exact lock-step, in *synchronisation*.

In those early days, facsimile machines managed to achieve good results using the precision of the vibration of a tuning fork. All that was needed was to line up the edge of the picture. At the start of each transmission, several seconds of what were called *phasing pulses* were sent. The operator manually lined up the receiver by slipping a clutch mechanism until the phasing pulses were in alignment. Later systems used that period of phasing pulses to make the adjustment automatically.

Whilst such a facsimile machine can be a fascinating instrument in its own right, the interest lies in the

Drum Scanning in Electronic Imaging

In 1848, Frederick Bakewell implemented the drum scanning approach used for the first facsimile machines. It was and remains a highly successful means of scanning and printing.

Drum scanning as described here is *still* used for scanning images, but is also used in the high precision professional scanning market where over 10,000 lines per inch are needed. (Standard computer flat bed scanners can achieve typically 1200 lines per inch).

Drum scanning has also been used to create large posters for advertising billboards, though wide format ink-jet printers have largely replaced it.

scanned, transmitted signal. That signal carries the brightness information of the entire picture represented as a continuous series of lines.

The mental leap here is thinking of the flat two-dimensional picture, in space, converted to a one-dimensional electrical signal, varying in time. The value of the signal at any instant corresponds to the brightness of a small spot in the picture. The time when that instant occurs can be measured from the start of the line and from the number of lines since the start of the scan. That overall time determines where on the picture the spot occurs (see Figure 5-2).

'France'   Union Jack

**'Fretwork'
Model 'B'
Televisor**

Fig 5-3. From the original picture of a prototype Model 'B' Televisor (top right), the reflection in the lens was enhanced (top left), compared with published material and discovered to be the scene at the International Industrial exhibition at Rotterdam in late September 1928. Below, the layout of the stand shows the Televisor on the extreme left. Next to it is the Televisor used by Clapp for demonstrations of long distance reception in Glasgow and the USA. Mid-stage is a stand portraying the newly-introduced 'Television' magazine. Further to the right is another 'fretwork' Televisor and finally a model of the 'double-8'. The original plate shows a dummy head with an extremely long nose.
Courtesy of the Royal Television Society RTS 37-46 (top) & RTS 36-91 (bottom)

Television Synchronisation

Television unravels the scene just like facsimile. The difference is that in television we repeatedly scan a continuously changing 'real-world' scene and hence generate a succession of new images. It is the repetition of scanning the scene that gives us the impression of capturing movement. At any one time, only a single brightness (or colour) value is being transmitted. The trick in television is to scan each of the images fast enough

Fig 5-4. Sydney Moseley looking at the Model 'C' Televisor – the Baird Dual Super Radio and Television Set, affectionately called the 'Noah's Ark'. This Televisor was first shown at Radiolympia in 1928. Photograph dated 30th September 1929.
Courtesy of the Royal Television Society RTS 36-82

that we do not notice the underlying structure.

Television gives us an extra dimension over and above that of facsimile. This extra dimension is the time separation between consecutive scans of the scene. Each scan, or TV frame, forms part of a continuous, never-ending sequence of frames. The frame number, or the time associated with it, is used in broadcast television today as an aid to editing together various scenes.

In general, timing is crucial to make a television system practical. If we think back to the two facsimile machines, a deliberate means of keeping them synchronised could be largely ignored. The early facsimile pictures only took a few minutes to be transmitted and were manually lined up at the start of each picture transmission. The tuning-fork arrangement was sufficiently stable that any skew in the picture caused by tiny differences in their respective frequencies at transmitter and receiver was minimal.

Fig 5-5. The production model of Televisor, which appeared in 1930.
Courtesy of the Author from Barton Chapple 'Television Today & Tomorrow', 1931

For the continuous reception of television images, we cannot rely on free-running both the camera transmitter and receiver display however precise they may be. We need a much more elaborate method of ensuring that we can show stable pictures at our receiver. The answer is to take an active approach to keep our system synchronised. We need something that will continually check and adjust the timing of our receiver against the received images from the camera. Somehow, we need the timing information from the camera alongside the images.

In early demonstrations of mechanically scanned television in the United States, the test broadcasts often included one channel for the video signal, a separate channel for the timing signal to drive the receiver and a third channel for the sound. Though it was impractical to use so many channels for one programme, the results were excellent.

30-line Vision Synchronisation

In Britain, the practical solution to a broadcast television service was one that combined timing with the vision signal. John Logie Baird developed an approach that was not especially reliable, but worked, after a fashion. He used no specific timing pulses and no separate timing channel, preferring to rely on the signal itself, which, being a television signal, generated a repeating pattern at the frequency of the lines.

Baird's 30-line television system needed the viewer to adjust the receiver motor speed until the picture was properly framed. Both the fixed structure of lines (set by the mechanical layout) and the Nipkow disc (or mirror drum) acting as a high inertia flywheel made an inherently stable system. Once the picture was aligned and stable, it would not need re-adjusting. So long as the timing for the lines was preserved, and so long as the motor ran at the correct speed, the picture would remain stable indefinitely.

Electronic Synchronisation

Electronic television from the middle of the 1930s onwards really could not do without explicit timing. Unlike mechanical systems, where the structure dictated where lines would appear, an electronic solution needed to be told where it should 'peg' the lines in place on the display screen. That meant each line and each picture (or frame) needed a unique peg to tell the electronics where the start of the line and picture were.

These 'pegs' or 'sync' (synchronising) pulses appear at the beginning of each television line and picture at a point in the waveform where there is no signal information. In order to discriminate between picture information and timing, the pulses are arranged to go outside the allowed signal range of voltage. The sync pulse drops below the 'black-level' voltage to what is essentially an illegal voltage value, making it easy to detect.

Today the concept of manually setting the start of picture and making occasional adjustments to keep the picture stable seem ludicrous. However, if we view that approach against what an operator had to do to run a facsimile receiver, it is roughly comparable. The analogy between receiving facsimile and receiving television is closer than we might expect. Just before the start of the BBC's experimental broadcasts of Baird's 30-line television in 1929, the BBC was broadcasting experimental facsimile images using the Fultograph system, described later.

In Baird's commercial receiver, the Televisor (see Figures 5-3, 5-4 and 5-5), the synchronisation of display with the camera in the studio relied solely on the repetitive structure of the incoming video signal. The idea was that the amplified video signal was not only fed to the neon display to generate the picture, it was also fed to two electromagnets placed either side of a set of multiple poles (see Figure 5-6). These soft-iron poles were arranged as one per TV line and mounted on the drive-motor shaft of the Nipkow display disc. If the disc started to drift slightly faster or slower, then the idea was that the electromagnet action would pull the disc back into alignment – against the turning effect of the motor. Though the rotating Nipkow disc or mirror drum was acting like a high inertia flywheel, the concept was that 375 (the line frequency for 30-line TV) tiny *pulls* per second over many seconds was adequate to correct minor drift. Work done by the BBC in the 1960s showed that the strength of the pull was inadequate. However, later work by an amateur, using an original Baird Televisor demonstrated that the method was quite stable over a few hours (more than the length of television broadcasts in the early 1930s) just so long as it was fed with a stable video signal.[1]

Whether the method of synchronising the Televisor worked or not, the concept of using the video signal – the picture content – to maintain a steady picture is decidedly unreliable. In the absence of an alternative, this however is the approach used for correcting the timing of the 30-line recordings.

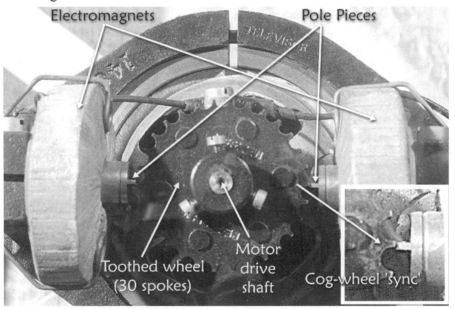

Fig 5-6. Detail of the synchronisation gear on a production Televisor.
Courtesy of the Author. Televisor from the NMPFT collection

What do we do without Timing?

The published specifications for Baird's 30-line standard showed that the signal had a short period at the end of the line, lasting only 3% of the line length, which was simply blanked off to give a value corresponding to black. So long as the rest of the image, or indeed the beginnings and ends of the lines, did not fade to black, the signal always then had a component at the line repetition rate.

Baird's Phonovision recordings have mostly a black background; there is no blanked area. This tells us that blanking had not been implemented at the time of Phonovision. Of course, Phonovision should not have needed timing embedded in the video signal. The concept for Phonovision was that the timing was to be inherent in the recording itself. Linking the turntable with a display should have resulted in a perfectly stable picture. None of the recordings achieved this, though the recording of Mabel Pounsford, the latest of the Phonovision discs, came closest. This means that for the restoration of all the Phonovision recordings we have only the video information with which to stabilise the picture.

In contrast, all the amateur recordings made of 30-line BBC TV broadcasts show an extended region that corresponded to a deliberately blanked-off area. This was however not the published specification of 3% but more like 10%. This much higher number gives the crude sync detector of Baird's Televisor display a more substantial signal on which to work. We may be witnessing an unannounced improvement to his standard, brought about after poor test performance with the 3% blanking period.

Early Attempts at Viewing the Recordings

Both the absence of timing information on any of the 30-line recordings and their poor state make restoration a serious problem. Before computer processing, all previous attempts had been done in *real-time* – on a live playback. These attempts had used analogue filtering to clean up the signal and elaborate hardware circuitry to try to synchronise the playback.

In the middle of the 1960s, engineers working at the BBC were successful in this approach, but only on one disc. They were able to grab fleeting images from the 1935 'Major Radiovision' disc of test stills (see Figure 5-7 (left)). At that time they did not fully appreciate where the 'Major Radiovision' disc had come from and believed that they were handling true Phonovision from Baird's time. They eventually realised that this disc was made at the end of the 30-line period.[2] Although they had succeeded in getting pictures, some amateurs also managed to retrieve pictures from the 'Major Radiovision' disc. Their efforts probably give the best pre-computer results (see Figure 5-7 (middle)).

When those same engineers turned their efforts to a Phonovision disc – the one held by the Royal Television Society (one of the recordings of

Fig 5-7. Various reproductions of a figure from the 'Major Radiovision' recording. On the left is the result by Tim Voore in the 1970s using analogue electronics and an oscilloscope for display. In the middle is the result from radio amateur, Chris Long. On the right is the author's attempt using computer processing.
Courtesy of Tim Voore, Chris Long and the Author

'Wally' Fowlkes) – they drew a blank. The approach of using audio filters (like a graphic equaliser), simple phase correctors and an oscilloscope for a display was just not sufficient for them to see anything recognisable on a true Phonovision disc. For them Phonovision remained an enigmatic curiosity and a good and solid example of one of Baird's failures. On only one occasion has a picture captured from a Phonovision disc using analogue filtering, but that bears almost no resemblance to what the computer helps us see (Figure 5-8).

Principles behind the Restoration Process

The computer for image processing is an ideal platform for delving into and restoring these recordings. In fact the computer, together with signal and image processing techniques and the ability to develop and tailor them for the purpose, forms a complete tool kit.

The general-purpose personal computer (PC) is not a perfect solution: it has features that make it more suitable to handle business applications and 3-D games than processing signals and images. Although computers can manipulate numbers very fast, they do have a finite performance and a limited capacity for processing. As the bulk of the processing is done in software, its execution will take longer the more complex the task is. Processing in real-time can be done, but for the 30-line vision recordings, there is absolutely no need to handle them in that way. Real-time processing

requires carefully constructed software, fully optimised for speed at the
expense of complexity. It is also difficult to do complex multi-stage iter-
ative processing, which is what these
recordings need.

The approach then becomes one of
capturing the video data into the
computer's memory, processing it off-
line, then displaying the result as a
cleaned-up television sequence. These
then are the three main phases that we
will now explore: Capture, Restoration and Display.

The **Capture Phase** takes the
original video recording on disc,
converts it to digital format and stores
it as a computer data file. In capturing
either part or the whole of a recording,
we are helped greatly by these
recordings being short – of the order

Fig 5-8. The only known off-screen image from a Phonovision disc (Anderson's RWT620-11) was captured (left) probably in the 1970s. The computer-restored image on the right came from the same disc.
Courtesy of the Author (right)

of a few minutes – and having no frequency higher than can be recorded on
a 78 rpm audio disc. Separating this capture and storage process has the
added benefit of giving us a reference set of unprocessed or raw data. We
can use this to assess how well the Restoration phase performs.

The **Restoration Phase** takes the computer data file and any collateral
information on how the discs were made, and analyses the content for
faults. Once the faults are detected, they are stored for analysis and re-
applied to the original data to cancel out the distortion they caused. Any
residual errors are in their turn analysed and the cycle of correction is
continued until we get an acceptable quality. The result is a computer data
file of the restored video recording. The stored faults are studied and
analysed for clues as to what caused them.

In the **Display Phase**, the restored video data file is processed for
showing in any format on any system: dedicated computer screen format,
standard computer video clip format, broadcast television or even in its
original 30-line format for showing on one of the few Televisors in
existence today.

Phase 1 – Capture

Baird did not document how the Phonovision recordings were made.

However we do know from the quality of the cut of the groove and the manufacture of the discs that they were made using the then latest professional recording techniques.[3] The subsequent amateur recordings onto aluminium were made using domestic recording equipment, notable more for its cheapness than its audio quality.

In order to capture the video signal from the discs, we need to convert the signal embedded in the walls of the groove on these discs into something a computer can handle. If the signal had been in some obscure standard, such as an obsolete videotape format, then playback would have created its own set of problems. As it happens, we are fortunate twice over. All television recordings were made on standard media for audio and the equipment to play back those old audio recordings is still around today.

"Say again, Houston? Find ancient images in this valley? You'll want me to look for diamonds next !!!"

D F McLean

Fig 5-9. A lunar rille, such as explored by the Apollo 15 crew in 1971, shows a passing resemblance to a groove in a gramophone record.

Courtesy of NASA and the Author

For playing back the signal, we need to convert the 'waggle' of the record groove directly into voltage changes in an electrical signal. As we are dealing with an analogue medium the process of getting from disc to computer has to be taken with some care. The first stage in the process is the transducer – the pick-up cartridge. This is the key mechanism that converts the path of the groove into an electrical signal.

The stylus on the cartridge must follow or *track* the groove correctly and precisely. In order to do so, the stylus has to have the best profile and shape for following the movement of the groove walls. Crudely, if the stylus is too narrow, it will run along the bottom of the groove and make poor contact with the walls. If the stylus is too wide, it will sit on top of the groove and track just the top lip. Ideally, the sides of the stylus should make the maximum contact with the walls of the groove, integrating out any small defects and surface irregularities. This can make a remarkable difference and the choice of stylus is down to the expertise of the person doing the transfers.

Pick-up Cartridge

A tiny stylus at the end of a thin metal rod fits into a cartridge, which is in turn mounted on a pick-up arm. The stylus follows the groove as the record is turned. Inside the *phono* cartridge, the movement of the stylus on the rod moves miniature coils of wire past magnets, inducing a tiny current. The induced current is the basic electrical signal that represents the path of the groove.

With the best possible signal coming from the cartridge, the signal has to be filtered, or 'equalised', to even out the frequency response of the whole recording and reproduction process. The most common equalisation is the transfer characteristic still in use today on the phono input of all hi-fi equipment. This, the RIAA (Recording Industry Association of America) characteristic, corrects for the late shellac (78 rpm) and all vinyl (45 rpm, $33^1/_3$ rpm) audio discs. It does this by applying a bass frequency boost and top frequency cut smoothly over all recordable frequencies.

For the Phonovision recordings made by the Columbia Graphophone Company, the RIAA characteristic is wrong. This is only apparent from studying the signal after restoration. The closest profile that gives the least distorted vision signal is a flat, even response at all frequencies, normally referred to as the 'Blumlein'

Synchronous Recording

In the 1980s when I first started to process the Phonovision recordings, home computers had only a tiny fraction of memory (and performance) of what we have today. In fact the storage capacity could only allow an absolute maximum of 36 consecutive 30-line TV frames to be stored at one time. Considering that a full Phonovision disc holds something like 750 frames, this was quite a challenge. It was not possible to load the entire recordings directly into the computer from the Phonovision disc on the turntable. I resorted to transcribing the discs using ¼ inch professional audio-tape as an intermediate format. Realising that tape itself could affect the timing, I modified my turntable to provide a fixed series of pulses on every revolution, recording these pulses on one track and the video signal on the other track.

As it turned out, the reel-to-reel audio recorder, a Revox A77, was far more stable in timing than the video of any of the discs that I transcribed. Consequently, the pulse track was never used (other than to prove this point).

characteristic. This was in use in the late 1920s by the Columbia Graphophone Company.

With the correct stylus, correct replay characteristic and a high quality conversion of the signal to a digital stream of data, almost all the physical world factors that affect the quality of the result have been dealt with. One physical task remains – the centring of the disc on the turntable.

Centring the Disc

In transcribing any gramophone, 'needle-in-the-groove' disc, one step that requires extreme care and attention is ensuring that it is perfectly centred. If the disc were to be slightly off-centre on playback, the pick-up arm would gently sway from side to side once every turn of the disc. The distance from the centre spindle would be varying with each turn of the disc. As a result, the instantaneous speed of the needle in the groove would also vary, speeding up as it swings outwards from the centre and slowing down as it swings inwards. For audio discs, the effect is a small but audible 'wow' in pitch on steady notes.

The effect of an off-centre disc on any of the 30-line video recordings is quite dramatic: the picture will roll one way and then the other on every turn of the disc. At first, it seems strange that a minor offset in placing the replay disc on the turntable can have such a major effect on the picture. However, we are relying on the playback being absolutely stable, as we have no timing on the video for creating synchronisation.

Just putting some numbers on it shows the problem. If we slip playback speed of a 30-line Baird standard image recorded at 78 rpm by just 0.25%, we will cause a displayed image to change from being stable to one that rolls once every second (that is, slipping by 1 line in 400 lines or 13 frames). In the same way, if the centring of the 30-line vision disc is out by just 0.5 mm, the Baird standard picture would roll three times one way then three times the other on every turn of the platter.

The effect on the early Phonovision discs, which have only three frames per revolution of the recording, is significantly less, but nevertheless still quite noticeable.

Digitisation

After taking all the physical steps necessary to ensure the best possible conditions for playing back the discs, we have the video signal prepared for the next stage of restoration. We need to convert it into a form that a computer can handle, and that means sampling and converting the smoothly varying voltage into a stream of numbers. This process is called digitisation.

Sampling the smoothly varying signal is just that – capturing the value of the voltage at regular intervals. The frequency at which we sample the signal has to be sufficiently high to collect enough samples to build up a picture. Too few samples and we miss information; too many and we waste memory storage.

Taking samples of the voltage at regular intervals gives us a sequence of stable voltage values that we feed to the converter hardware. Each stable voltage value is converted into a number, represented in binary notation to reflect the hardware implementation. The scale of these numbers is adjusted so that the extreme numeric range represents the extreme range of brightness values. For an 8-bit wide binary number, those extremes are 0 and 255, equivalent in binary notation to 00000000 and 11111111 respectively. Conventionally, these represent the levels for black and white respectively. The stream of numbers is created into a list of values that are stored in the computer as a data file holding the raw, unprocessed data. The signal is now digital and is the starting point for digital signal and image processing.

The sampling and conversion stages, normally called Analogue-to-Digital Conversion, are usually carried out in a single device. Today that device is often found as a standard item in sound cards for computers. In general, these mass-produced items fall short of professional quality and may in themselves add to the distortion.

The frequency for sampling a signal should be a minimum of at least twice the maximum frequency within that signal. This minimum sampling frequency is called the Nyquist frequency, named after the mathematician who proved that such a waveform could be represented perfectly as a series of numbers. For Phonovision, the home computer technology of the early 1980s initially limited the maximum sampling rate to 64 samples per line, though easily meeting the Nyquist criterion.

With the rapid improvement in computer performance and storage capacity, the sampling rate was raised to 140 samples per line. Equivalent to 52,500 samples per second for the 30-line Baird standard, this high sampling rate is not strictly necessary to capture all the video content. It does however allow greater precision in correcting the timing of the recording, giving rise to a better quality result.

Phase 2 – Restoration

Capturing the video recordings into the computer and storing them as standard computer data files is a major step along the path of their restoration. The whole gamut of computer-based techniques is available for us to explore, adapt and apply to the task. The challenge then becomes one of developing and enhancing those techniques to be used in a way that they have never been used before: restoration of television recorded on audio discs.

Most of us today expect to be able to buy an off-the-shelf sound processing package or a graphics 'paint' package and use the signal and image processing functions already installed. These pre-packaged functions are designed for processing audio as sound, imagery as scanned photographs and video as compressed versions of current broadcast television standards. Not one of these is usable for the task in hand.

The 30-line video recordings contain a non-standard obsolete video format. They were recorded on a medium not designed for video, have experienced faults made at the time of recording and have degraded over the seventy years age of the medium. This all means that we have an entirely unique set of faults requiring a completely custom approach to tackle them. This custom nature does need emphasising, and is the single most important reason why no one else has undertaken this work. The investment in time handcrafting the software, trying out technique after technique and painstakingly restoring the images is immensely labour-intensive. This is not helped by the distortion being excessive at times and being different virtually from one disc to the next.

Carrying the responsibility for bringing to light the earliest-known recordings of television means there have to be some ground rules for their restoration. We must first try to get close to the original scene quality. In attempting to do so, we must neither remove anything of value, nor add anything that was not there to start with.

With today's sophisticated graphics systems, it is so easy to go beyond the restoration process into the realm of simulation. Simulation is an important and powerful tool for illustration. When it comes to enhancing the material though, the task is one of clearing away the distortion, leaving behind a signal that is the best representation of what the original vision signal was like all those decades ago.

In order to achieve this, we need to select techniques that will allow us to identify and measure the distortion, and then develop techniques for suppressing and removing the distortion, thereby restoring the signal.

Signal Processing

Digital signal processing is the generic set of computer-based tools for handling signal waveforms. In fact, it is a complete subject area in its own right that occupies many volumes of learned wisdom.[4] Signal processing techniques are used in all walks of life. We find them in mobile and

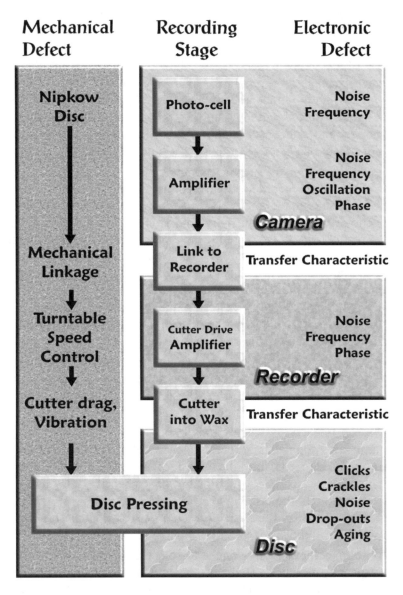

Fig 5-10. The sources of distortion categorised by mechanical, electrical and functional stage in recording. The camera system, the recorder and the disc itself all have their own distinct sources of distortion.

Courtesy of the Author

satellite phones where they allow the most efficient use for voice communications. They are also used in speech recognition, in music synthesisers and sound processors, and in all radar and sonar systems. In fact, a signal from any type of measuring and sensing device is almost always subjected to some form of signal processing.

The file of 30-line digitised video data from the Capture Phase represents closely the signal waveform directly from the discs. We can use the digital signal processing techniques to begin our understanding of just what the real problems are (see Figure 5-10).

The signal recorded on the discs started out when the light reflected from the scene was picked up by the photo-detectors. The detectors had a particular characteristic that altered the signal. The signal was modified yet again by the electron-tube amplifier that boosted the signal.

For Phonovision, this amplified signal was fed through some link – either via a transformer coupling or even across a loudspeaker-microphone air gap – to the recording equipment. The signal then passed through the front end of the recording equipment and then to the disc cutter. All these stages contribute their own signature of added noise, limited frequency response and shifts in phase response. The disc cutter carved out a miniature line graph of the waveform into the warm soft wax surface of the master. From this master the pressings were made preserving the signal until they were replayed decades later and captured into the computer. This is the complete signal path encompassing a whole variety of signal distortions, many of which we are only able to guess at.

A similar process occurs for the later amateur recordings. We have the additional steps of processing the signal in the studio, transmitting the vision signal, receiving it and recording it onto the domestic aluminium disc-blank recorder. In parallel with the sources of electronic defects, there is another set of defects arising from the mechanical linkage. These defects tend to affect only the video timing.

The electrical signal that goes through all these possible stages of degradation is of course television and has the structure and timing that accompanies it. Unlike a pure audio waveform, the structure of television means we have a regular sequence of adjacent lines spaced at equal intervals.

This is where we can use the line-to-line and frame-to-frame relationships of the recorded images to help in restoring the signal. This relationship means we can develop suitable digital image processing techniques to understand the problems and to apply corrective measures.

Image Processing

Digital image processing was a relatively new subject when this work started in the early 1980s. It had grown out of the need to correct for faults in images taken by space probes, starting in the 1960s. Those faults showed up as artefacts on the imagery and were caused by errors in the camera and in the communication of the image back to Earth.

Before we can apply any digital image processing techniques to help in the restoration, we need to convert the list of numbers representing the digitised signal into an array of numbers, representing images. This is where we start to depart radically from the academic textbook techniques for digital image processing.

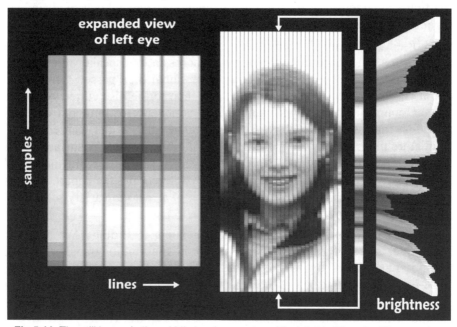

Fig 5-11. The still image in the middle has been reduced in detail to 30 vertical lines each of 140 separate samples. This digitised image is clearer on the zoomed image on the left, showing 8 lines of 24 samples per line. Taking one of the scan lines, we can plot the progression of brightness along any one line (right), demonstrating that we can consider a TV picture as either an array of pixels (left, or as a continuous sequence of waveforms (right)

Courtesy of the Author

'Classic' textbook image processing techniques assume that we have a picture, or a photograph, divided into a grid of squares (see Figure 5-11).[5] This is similar to the sampling described earlier for signal processing but extended in two dimensions to cover the picture. The brightness (and colour) values in each of these squares is averaged out and converted to one or more numbers. The terminology for each of these squares is *pixel*, a lazy version of 'picture element'. If we convert a colour image, we might have three values per pixel, one value for each of the colour components.

To get photographs stored into a computer, the scanner will convert the photograph into a grid of pixels, where the coarseness of the grid will determine how much detail we see on the eventual stored picture. For television, each TV frame or picture is already scanned into a stack of lines. The line structure then forms one scan path and we only need to slice up each line into the correct number of samples per line to complete our grid of pixels. This means that the process of capturing a television signal has more to do with signal than image processing (see Figure 5-11).

For the material on these recordings, where we have the one-dimensional scanned signal distorted (rather than the two-dimensional image), we are far removed from simple, classic image processing problems. This directs us to focusing on a hybrid of signal processing (along the lines) but using the fact that we have images here to assist with the restoration.

We can attribute the distortion to a series of faults, most of which have been caused by separate problems and can be treated as independent of each other. We can classify them into a few categories and then look at the reasons for the distortion. The categories for the various distortions are Timing, Frequency/Phase Response, and Disc Defects.

It's all a Question of Timing

Of all the faults on the discs, the worst type by far is in the timing. Each of the disc recordings has one or more different timing faults that seriously distort the picture. Those timing faults are the single reason that the images on the discs have remained unseen since the time they were made.

Fundamentally, the timing of the video signal replayed from a disc is not stable. The lines and frames occur at slightly irregular times. The display system – the Televisor – cannot cope with the variation. The only way to tackle it is to fix the problem at source, by correcting the timing of the video signal.

The original 30-line television camera for these recordings was a source of inherently stable video. If we made a perfectly stable recording, the absence of any special timing signal for synchronisation would simply result in a picture that would 'hunt' or gently roll a line at a time. The only recording that manages to achieve this level of stability is the most recent 30-line recording, the commercial disc of test stills – the 'Major Radiovision' recording.

Without timing information, how do we know what the errors are? These recordings were made of a video signal without any timing embedded in it. We can, however, exploit the picture information. If enough pictures are taken every second of a subject that hardly moves, then the picture should be much the same from one frame to the next. We should be able to match up the pattern of the picture across several TV frames, find out how much it has drifted by, and re-correct the timing so that the pictures line up on consecutive TV frames.

This is great in theory, but in practice almost every stumbling block is in our path. Each disc has its own problems, but let us take the earliest Phonovision recording as about the worst example of what we have to work on.

- First, the recording speed is changing fast, in less than the time to scan a single TV frame. The simple approach of matching up complete successive pictures just will not work reliably as the pictures will be heavily skewed and distorted.

- Second, the picture repetition (frame) rate is low. The picture can change quite radically in appearance on successive TV frames, spoiling attempts at matching up common features.

- Third, the picture has no low or high frequencies. If we plot the brightness of the image along one of the TV lines, we see, instead of crisp and clear details, nothing more than a sine wave varying in amplitude. When we look at the picture though, we see a face. This is more to do with the power of interpretation by the brain, looking at the movement and actions over time than what is *really* there in the analytical sense.

- Fourth, the detail in a picture, built up from only 30 lines, does not give a significant amount of information on which to create a match.

Just establishing these points and their relative importance took some time and a great deal of experimentation. The starting point was to match lines and frames and from that, understand what had to be done to tackle each of the problems. At the core of the approach is a simple mathematical technique – pattern matching.[6]

Pattern Matching

Matching lines with lines and frames with frames is the first stage in correcting the timing. We start by assuming that each successive line and frame should be in their correct position and then search either side of that position until there is a match. This stage tells us how far each line and frame is out of position, allowing us to move those lines back.

We find the best match by sliding a copy of the pattern that we want to find across the uncorrected data (see Figure 5-12). A few techniques can be used to find the match. Both take a common approach.

First, a block of data that is correctly positioned is stored as a reference. This can be a TV line, frame or some other pattern.

Second, the next available block of data from the uncorrected part of the data is matched against the reference data over a range either side of where we expect to find a match. At each position, the degree of match is calculated. The position that creates the best score tells us where the pattern appears in the data.

Third, the block is shifted back by the amount worked out in the matching process. This effectively lines it up with the video data that has already been correctly positioned.

Fourth and last, the reference data is updated to reflect the values coming out of the matching process. This allows changes to the pattern, caused by movement of the subject, to be incorporated.

Pattern Matching Methods

The two techniques for pattern matching are *correlation* and *sum-of-difference* (Figure 5-12). For correlation matching, the brightness value of each point in the reference is multiplied by the corresponding point's value in the uncorrected block of data. The results of all these multiplications are added together, creating the correlation score. A score value is calculated for each of the shifted positions of the uncorrected block. The position with the highest score is where the best match occurred.

The second technique is almost identical, except that instead of multiplying corresponding values together, they are subtracted. The sum of all the absolute differences (hence sum-of-difference) is the score value. The best match occurs when the score value reaches a *minimum*.

Though correlation matching is much more robust than the sum-of-difference technique, it can also be lengthy on computer time. To correlate one 140 sample per line, 30-line TV frame within a range of two frames in the source data would require 35,280,000 unique multiply and add operations. In practice we do not need to resort to such a wasteful exercise but there is no escaping the intense nature of numerical calculations for doing this matching automatically. For this reason, the original computer used for processing Baird standard video incorporated a high-speed hardware multiplier chip, as those very first microprocessors were particularly slow on calculations involving multiplying and dividing.[7]

The first crude automatic attempt at correcting a sequence of 32 frames in a Zilog Z80A based home-built computer took 96 minutes in assembler

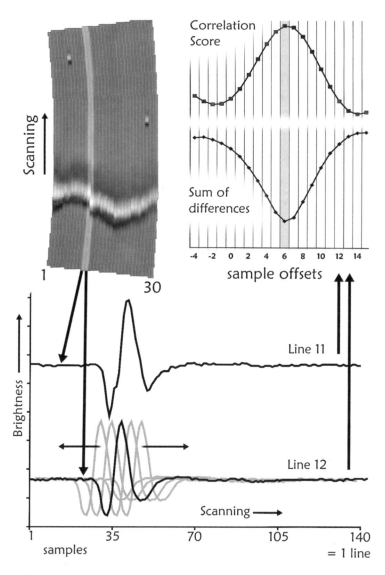

Fig 5-12. An example of the two methods of pattern matching. The graph (bottom) shows the two highlighted adjacent lines, 11 and 12, from an uncorrected image at the start of the earliest known Phonovision disc, SWT515-4, 20th September 1927. Line 12 is moved one sample at a time either side of its original position. At each position a value is generated by multiplying (correlation score) or subtracting (Sum of Difference) each corresponding value. All the (absolute) values are added together to create a score. This is plotted for each offset position of line 12 in the graph at top right. The best estimate is that line 12 is 6 samples (out of 140) out of position.

Courtesy of the Author

language. With the same calculations in compiled PASCAL computer language, the time was 17 hours. For interpretive BASIC, the time would have been around 75 hours.

Matching the current TV line against a history of previously matched lines was a technique developed in 1981 for Slow-Scan Television (SSTV).[8] The SSTV images were an excellent proving ground for the matching technique and showed that the technique could be used to correct the early vision recordings.

Phase and Frequency

There is a feature unique to video that was understood, but none too well, back at the time of these recordings. Even today, none but the broadcast television specialists need worry too much about it. This feature is the effect caused by phase errors. It arises primarily where there is a poor low frequency response.

Generally, audio recorded onto shellac, as in 78 rpm discs, suffers from having the very low frequencies and the very high frequencies cut-off. The absence of low frequencies makes the sound lack punch and the lack of high frequencies makes the sound muffled. Columbia's Western Electric system, first introduced in the United States in 1925, had a

Fig 5-13. The test image on the left and the picture of Edward, Prince of Wales on the right both illustrate the effect of poor phase response on vertically scanned 30-line video. On the left, the effect is the light shading above the black diagonals and the white flare above the vertical black line. On the right, the effect shows up as white streaks above the dark hair.
From originals courtesy of D. R. Campbell

frequency response of approximately 50 Hz to 5,000 Hz. To our ears, the sound appears 'recorded' with the added signature unique to *needle-in-the-groove* recordings of clicks, crackles and surface noise.

By recording 30-line video onto wax discs in the same manner as for audio, we get those same degradations. However they have a far more marked effect on video than audio. The lack of high frequencies makes sharp edges in the picture appear rounded. The lack of low frequencies is more complex. We see strange shading along the lines: bright objects get progressively darker, dark regions get brighter. This shading is not simply due to the lack of low frequencies but arises from shifts in phase of those low frequencies (see Figure 5-13). Such phase errors can be corrected either digitally or in the analogue domain. The correction does not restore the low frequencies, merely the phase relationship of the existing frequencies.

In audio, our ears cannot detect small phase errors, yet their results show up clearly on video. This difference between how our ears treat audio and how our eyes analyse video highlights the major difference between audio and video, and it really comes down to how we process the information. Video is a complex waveform where the value of each point in the waveform needs to be a faithful representation of the brightness in the scene. We see it discerningly, analytically, plotted almost like a graph on our screens. If we plotted audio on an oscilloscope, we would see effects, like phase errors, that we were unable to hear. But our hearing is not analytical; it is interpretative.

Whilst we enjoy and prefer to listen to music in high fidelity audio, we can still understand music, speech and other sounds in quite poor conditions. The early cylinder recordings, or disc recordings recorded acoustically, are quite intelligible though low in quality. Our ear provides the brain with the raw information on loudness and frequency. Our brain then does the hard part in elaborate processing to convert the data into recognisable sounds. Part of what helped our ancestors defend themselves and hunt, allows us now to enjoy the great composers, jazz, blues or rock. That ability makes us enjoy the music irrespective of it coming from a cheap tinny-sounding radio at one end of a large reverberant room, or performed live in concert-hall conditions. It allows us to understand one person speaking amongst others at a cocktail party, and hear our name mentioned amongst the hubbub.

Human vision is both analytic and interpretative. When we cannot see an object clearly, or when there is noise or distortion, our eyes and our brains will attempt to understand what we see. The amateur recordings of 30-line BBC Transmissions have a high degree of phase and frequency errors, part of which come from the recording process and part in the reception of the signal. How much of each is difficult to say. The Phonovision discs lack most of their low and high frequencies. Yet we see the recorded subject in all of these quite clearly. Even though we should not see anything, our brains recognise the movement as being human and can interpret it as, for instance, a troupe of dancing girls.

Fourier Analysis

The technique to study the problems in phase and frequency is Fourier Analysis. This mathematical technique allows any complex waveform with a repeating pattern to be thought of as being built up from a series of discrete sine waves, each a harmonic at a multiple of the frequency of the waveform. The pure tone of a whistle is a sine wave, as is the mains hum that can be heard on poorly connected audio equipment. Any complex repeating waveform is made up from component sine waves (the harmonics) all with different features. Those features are amplitude, or the relative sizes of the sine waves, frequency, or how rapidly each of the sine waves change, and phase, or where each of the sine waves start in their cycle at the beginning of the waveform.

Figure 5-14 illustrates how we can create a simple square wave by adding the component parts – the odd harmonics – together in appropriate fractions. A square wave is a special case as it is so simple to create. The perfect square wave includes an infinite sequence of harmonics. In reality, any electrical system, whether it is an amplifier or even just a length of cable, has upper and lower limits on its frequency response. If we only include a few of the harmonics, to create our complex signal, we get the rippling effect as shown in the graph. We see this rippling or 'ringing' effect on all the 30-line video recordings. The size, frequency and phase of the ripple tell us what the highest frequency on the recordings are and what is happening to the other frequencies.

This square wave is quite a useful analogy to the white bar on the earliest Phonovision recording. It looks like a white bar if the brightness is plotted increasing vertically on a graph – high values being white and low values being black (see Figure 5-12). The illustrations of phase errors show

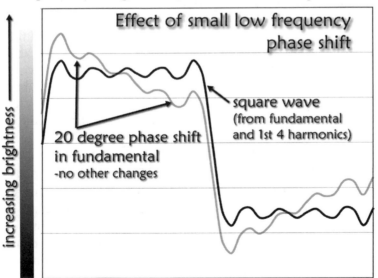

Fig 5-14. A simulated square wave built from the first four harmonics and the effect of shifting the phase of fundamental frequency by 20 degrees (out of 360). The amplitude of all frequency components remains unaltered.

Courtesy of the Author

the strange streaking effect away from rapid, large, changes in brightness. In our Fourier simulation of the square wave we can very easily simulate this effect. Figure 5-14 also shows the same square wave as earlier but with the lowest frequency component suffering a shift in its phase. It comes as quite a surprise to find that both these waveforms in Figure 5-14 have exactly the same frequency response. Simply shifting where the lowest sine wave component started – what we call its phase angle – by only 20 degrees brings about this massive distortion. The tilt in the waveform would equate to a gradation in brightness, which is exactly what we see on the video recordings.

Correcting for Disc Surface Faults

Nostalgia has become big business and nowhere more so than in the music industry. The advent of digital audio systems based on the Compact Disc created a new medium for re-issuing historic recordings. Gramophone recordings of major artists, mastered on tape and sold on vinyl from the 1950s, were re-issued on the robust, noise-free new medium. When it came to the earlier material, before the advent of magnetic tape, the scratches and surface noise of the masters, on both shellac 78s and the earlier wax cylinders, were reproduced faithfully by the new digital medium. There needed to be some way of eliminating this 'signature' of the gramophone.

Digital audio restoration is now central to bringing the sound of yesteryear into our digital hi-fi systems. The techniques have been around for some time and new approaches to improving the performance and quality are still being developed. The objective is clear: convert those old noisy recordings into something that is more attractive to the ear. The theoretical ideal is to separate the problems of the gramophone master from the musical content. If we were to achieve that, the results would be almost like taking a time machine back to the time of the recording and listening to the performance. Some restorers, like Ray Parker, have produced results of subjectively stunning quality without resorting to advanced digital techniques. Others have applied proprietary defect-reduction systems such as CEDAR (Computer Enhanced Digital Audio Restoration).[9]

All these approaches are pleasing to the untrained ear but fail to satisfy the purist. There is an argument that the reduction of defects is merely cosmetic and could detract from the musical content of the original recording.[10] That approach is more valid for extremely rare material where maybe only a single pressing remains. The transcription to Compact Disc then becomes more of a reference, almost archival copy. Whether they use advanced software or traditional restoration, all these approaches have one thing in common: they use signal processing to remove *clicks* and to suppress various degrees of defect reduction.

The challenge for audio gramophone discs is to be able to discriminate between the disc content and the generic noise of the medium. Clicks, or disc defects are usually detected by looking for a step shift in the playback stylus's movement. The stylus on the pick-up arm will normally follow the smooth movement of the recorded material in the groove. A crack or piece of dust will cause the stylus to move much faster (that is at a higher frequency) than the sound recorded in the groove. Natural sounds may have a sharp 'attack' (like a crash of cymbals) but rarely have a sharp cut-off. As a result, the defect detector usually uses a sharp cut-off as an indication of a defect. Once the system has detected a defect it joins up the audio either side of the click, effectively removing it from the audio.

Early Videodisc Defects

How can these techniques be used for the videodiscs? They were made using the same gramophone process as the 78 rpm disc. Consequently they suffer from the same audio faults: clicks, crackles and surface noise. The simple answer is that off-the-shelf audio signal processing software and hardware must not be used. The nature of the video signal is that it is usually at full amplitude with plenty of 'black' and 'white' objects. There are no quiet passages in video. The vision signal is quite unnatural as audio: it has rapid transitions that might be detected as defects. Most importantly though, any audio hardware that processes the signal will be useless for video as they alter the low frequency phase response in an uncontrolled manner. We have to develop new methods of removing these disc defects.

In general, the quality of the Phonovision pressings is high but there is nevertheless some residual noise. The later amateur videodisc recordings were cut into aluminium blanks. They not only suffer from similar surface noise problems to the shellac Phonovision discs, but have additionally been quite badly damaged from the build-up of corrosion of the aluminium over the decades. This gives rise to massive drop-outs of information where several lines or even several frames at a time can occasionally be lost and drop-outs from scratches can be as high as one hundred per frame. For a 30-line picture, this is exceptionally poor quality and seriously degrades what we see (Figure 5-15). As this is simply a bigger problem rather than a different problem, we can use the same techniques on both Phonovision and the later amateur recordings.

There are many advanced methods for restoring audio. They mostly rely on a statistical model to provide the discrimination between what is audio and what is defect. Their approach relies on classifying the defect and correcting for it, but very importantly involving a model of what is acceptable and detectable to the human ear. Going a stage too far can risk creating audible correction artefacts. In any case, digital audio restoration techniques are not directly appropriate to the problem. Their use is based largely on a subjective assessment of audio quality and the common use of audio masking techniques that are entirely inappropriate for a video signal.

Video has a distinct advantage over audio. The television signal has an underlying structure of lines and frames. We would expect to see the imagery generate repeating patterns that fit this structure. We can look for those repeating patterns by comparing the statistics of the signal along a line, across adjacent lines and at the same line across adjacent frames. This exploitation of the video structure across lines and frames provides the best method of determining automatically what is signal and what is noise.

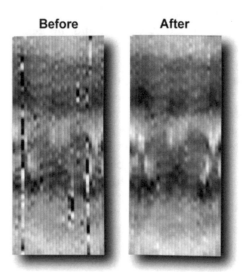

Before   After

Drop-out & Noise Reduction

Fig 5-15. One frame from a 1933 amateur recording made on an aluminium disc, illustrating the effect of automatic noise reduction tuned to the nature of the 30-line television image.
Courtesy of the Author

The principle is quite straightforward and is widespread in digital TV noise reduction systems. The method relies on the object being televised as either stationary or moving smoothly. In that way, the object will appear on adjacent TV frames following the path of movement. Drop-outs will however appear randomly and should bear no relation to the picture.

Whereas for today's television imagery there is sufficient detail to work with, the Baird 30-line vision signal presents a serious problem. The low detail and low picture refresh (frame) rate (especially for Phonovision) means that the object can move and change shape between successive frames by quite a large amount. Proportionately, noise and drop-outs take up a much larger portion of the frame than for today's television picture.

The trick in this approach is of course to ensure that the timing has already been corrected and that all the TV frames are perfectly lined-up. The trouble is that timing correction is sensitive to noise and drop-outs in the picture. This is normally the best place to start for audio restoration, as there is no way of detecting defects automatically other than by their statistics. To avoid getting stuck in a loop (to reduce the noise we fix the timing, but to fix the timing we must reduce the noise…), the method for correcting the timing errors has to be made as robust as possible. This has taken the bulk of the software development work. However, getting this right has been the key to successful restoration.

Filtering

When it comes to suppressing clicks, drop-outs and noise, the approach is one of trying to isolate them. For that we need at least three points whose values we can filter. We use three points on the assumption that in any three, only one will be 'bad'. In addition, the low resolution of the 30-line system and the low picture refresh (frame) rate means that three adjacent samples are just about the maximum we can reliably use. These three samples can be taken across lines and across frames. For a point, say partway along line 11 of one frame of the recording, we pick out the same point on the lines either side, on lines 10 and 12. We take those three brightness values and filter them. For filtering across frames, we need to take into account that enough time has passed for an object to have moved between frames.

Filtering along lines is done by a variant of the audio disc defect detection implemented in software modified and customised for video. The approach detects clicks and drop-outs up to the length of one line. Once the defect is detected, it is filtered using adjacent lines and adjacent frames on the same line.

The mean and the median filters are simple, yet extremely powerful. The mean filter calculates the average value and the median selects the middle value of the ordered list of numbers. If we have three values – 10, 255, 15 – the output of an averaging filter would be 93 and of a median filter, 15. The median can completely remove single isolated defects. It is somewhat fierce and can also have unusual and unnatural looking side effects.

Those side effects can be a little too effective. In support of a patent application on removing film defects and blemishes, I subjected a sequence of frames from a wartime newsreel of a Nazi conference to a technique based on the median filter across adjacent frames. The original had many blemishes, but also featured flash bulbs popping every few seconds. After the filter, not only had the blemishes been suppressed, but also all the camera flashes were completely removed, subtly changing history.

Fig 5-16. Filtering across frames is done by searching for a three-by-three region that looks the same on the previous and next frames. In the illustration above, the singer has moved her head slowly to the right by one region each frame. Once this movement has been detected, the filtering is performed along the path through the three adjacent frames.

Courtesy of the Author

Small Defect Detection

The way we locate defects in gramophone video is loosely based on that for audio discs. However in order to discriminate between defects and the video image, we use the relationship of the image across neighbouring lines and frames to help in detection. Whilst the defect will occur randomly, the image will have a structure across the lines that will change from one frame to the next as the subject moves.

The technique uses a small patch around the point being processed in order to get some structure to the current image. It then examines the image before and the image after, looking for all possible movements of the patch between images and selects the most likely path of motion. Compensated for the movement of the object in the image, it should see only a gradual change across the images. Any defect will show up on only one of the image patches. If it finds a sufficiently large error it will remove it using either the average or the median filter described earlier. The technique is called 'motion-compensated time-domain filtering' as it takes account of any movement of the object – such as a head moving from side-to-side (see Figure 5-16).[11]

Alternative Approaches

There is a wealth of alternative techniques for restoration that could be applied to these videodiscs. The problem with those techniques is that they were designed for images that are more conventional. They assume a certain level of detail and quality that simply is not there on the 30-line material.

Some of the more advanced restoration techniques require precise knowledge of the original camera and recorder. None of this information is available for the *gramophone videodiscs*. We simply do not know the precise circumstances of when the recordings were made, how they were made, and what exactly was used to make them (see Figure 5-17). On top of that, we have multiple faults in the recordings that affect the signal and the timing.

Fig 5-17. The 'Major Radiovision' disc suffers from a 5 kHz resonance which mars the video quality. On the right, the resonance has been removed and the arc-scan geometry added.
Courtesy of the Author

Whilst there are ways of correcting signal errors, timing is the one fundamental element for television that is essential to re-construct the image. All of this effectively directed effort away from classic image restoration and towards the more exploratory and investigative approach described in this book.

Phase 3 – Displaying the results

The corrected results of all the different stages of processing from Phase 2 are held in a computer data file. From the data, we can generate from the computer any type of video that we wish. However we have to be careful, as it is so easy to create more artefacts by displaying the data in the wrong way.

The gramophone videodiscs are recordings of a 30-line TV camera system that captured images differently from broadcast TV cameras of today. In conventional television cameras, the image from the lens is integrated or exposed for the time it takes to scan one frame. As each frame is read out sequentially from the camera as part of a continuous sequence, the next frame is being exposed onto the photosensitive surface of the device. At the end of each frame scan, the process repeats. This integrating principle is the same today in our solid-state chip cameras as it was right back at the very first electronic cameras of the 1930s (excepting the Farnsworth image dissector). We can largely treat each of the frames from such a camera as if it were an individual photograph.

This is not so with mechanically scanned cameras. For both types, the Nipkow disc and the mirror drum, only the single point on the image currently being transmitted is exposed. Those cameras do not build up the image. Consequently, there is never any motion blur as the instantaneous exposure is of the order of one ten thousandth of a second. Each of the frames generated in this way is a progressive series of short exposures spaced evenly in time. For instance, the 1st line represents a time that is 29 line-times before the 30th line. As a result, each mechanically scanned frame cannot be treated as an individual snapshot. A focal-plane shutter on a single-lens reflex camera gives a similar effect on fast-moving objects.

For a part of the Phonovision recording of Baird's 'Stookie Bill' dummy head, the head is rocked from side to side. Baird's 30-line system had vertically scanned lines, so the scan of each frame sweeps horizontally across the scene. If we treat each frame as a single still picture, we will see something strange happen. With the head rocked in the direction of scanning the picture, we will see the head stretched wide. With the head moving against the picture scanning direction, the head will be compressed in width. This is an artefact of the mechanical scanning made worse by the low frame rate of the Phonovision recordings.

As the restoration is done in a computer, we would expect to display the result on the computer's graphics display. This gives single frames of excellent quality, but lacks realism. The problem is in accurately simulating a single spot tracing out the image on a display, which itself is refreshed in its own television format. We get a massive interference or *beat* pattern between the frame rate of the 30-line television image and the frame rate of the graphics display.

The most realistic television image comes from viewing it on a display that matched the scanning of the original camera. We can use either an oscilloscope display or an original Baird Televisor. The key display feature is that it should have no 'memory' or storage. Phonovision shown in such a way gives the most remarkable result of all the videodiscs. The faces appear ghost-like, with an overwhelming sense of life and presence. The image though is distinctly ephemeral, because it cannot be filmed or captured onto videotape without tremendous flicker.

That flicker is the effect of the 30-line frame rate interfering or beating with the frame rate of the broadcast camera and has affected every attempt to capture such television images. In fact, when TV journalists say that 30-

line television had excessive flicker, they usually illustrate it quite incorrectly by showing this very beat pattern.

For displaying the video images on a computer, we need to convert the video signal into a sequence of separate pictures, which can be shown 'flick-book' fashion. This completely avoids the beat pattern but adds the unnatural appearance of a sequence of static pictures (see Figure 6-15 (top row)).

Vision restored

The restoration story is akin to that of handling archaeological discoveries. There, the artefacts are restored based on what they appear to be – pottery shards, jewellery, coins – and where they were found – a peat-bog, on the sea-floor or in someone's back garden. Similarly, the gramophone videodiscs require restoration based on which type they are – Phonovision, recordings of BBC transmissions and the commercial test discs – and what equipment was used to record them.

The tools to unearth and restore the video recordings and to understand the circumstances surrounding them have had to be handcrafted and adapted. In this book, it has only been possible to touch on the methods used. They however illustrate just what we should be prepared to do in order to preserve and restore our technological heritage.

[1] SMITH, P.: Private communication, 1999

[2] BRIDGEWATER, T. H.: Private communication, 1983

[3] LEVIN, E. B.: Private communication, 1996

[4] Refer to Bibliography for Signal Processing

[5] PRATT, W. K.: 'Digital Image Processing', 1981

[6] MCLEAN, D. F.: 'Using a Micro to process 30-line Baird television recordings', *Wireless World*, Oct 1983, pp66–70

[7] WAYNE, G.: 'Pixels from the Past', *ST World*, May 1990, pp23–26

[8] MCLEAN, D.F.: Letter on computer-controlled slow-scan television, *RadCom* (RSGB), Oct 1982, p872

[9] GODSILL S. J. & RAYNER P. J. W.: 'Digital Audio Restoration' (Springer-Verlag), 1998

[10] LEVIN, E. B.: Private communication, 1998

[11] HUANG, T. S.: 'Image Sequence Analysis' (Springer-Verlag), 1981, Chapter 4

6   Discoveries

'"We are now planning", he began, "a vast programme of research to extract all available knowledge from the record."'

On discovering a reel of film as the last remnant of civilisation on Earth.

A. C. Clarke, 'Expedition to Earth', 1954

Phonovision: Genuine or Fake?

From time to time, I am asked to investigate some newly found video recording supposedly made by Baird. It is quite natural for us to hope for the best, for a link with Baird, whether for romantic reasons or otherwise. Quite often, the only copy of the recording is on audiotape, leaving no possibility of checking the original material. Considering 30-line television finished in 1935, a recording of something that sounds like the 30-line vision signal ought to be from before that date. The problem is that several amateurs since 1935 have built equipment that generates low-definition television.

One such amateur, Bill Elliott, a Granada TV engineer, made 30-line TV recordings in the 1960s. He had two original Baird Televisors, one of which he converted to being a 30-line camera. One undocumented recording fooled a number of people until it was realised that Elliott had televised a replica of one of Baird's 'Stookie Bill' dummy heads. The give-away was the pristine quality of the tape recording (clearly done on a modern recorder) and the presence of large electronic sync pulses within the video signal. There was of course no intention to create a fake.

Prior to the 1980s, no one had been promoting the Phonovision discs as being of any value, historic or monetary. Their content was unknown and their significance not realised. As such they were unlikely to be deliberate fakes. Nevertheless, I needed to determine whether the discs were genuine or not, rather than leave any question of their authenticity open.

Fortunately, at least one of the discs is traceable through paperwork right back to 1928. This is the disc that Baird had donated to the Science Museum, as recorded via its Museum catalogue number. Baird had also donated another Phonovision disc to the Television Society. However, the Society, now sporting a Royal prefix, had not recorded evidence of the disc's pedigree. Indeed for many years, an early notary of the Television Society, Barton Chapple, had kept the disc and other precious items in safe keeping away from Central London and the threat of German bombs during the Second World War.

For the remaining Phonovision discs, there was no evidence, other than their great similarity in content and recording identification. Scratched onto the wax surface and preserved on the pressings were reference numbers and initials, possibly of the recording engineer, as shown in Table 6-1 below.

Label	Date on Label	Disc Surface writing
Test Record SWT515-4 Television 20/9/27 515-4-GS. Columbia Graphophone Company, Ltd	Stamp: 20 Sep 1927	
Test Record RWT620-4 Television. Not good. AP (or HP) 620-4 Columbia Graphophone Company, Ltd	Red ink stamp: 10 Jan 1928	#4 RWT620-4 EC #3 Wally
(No Label)	No date	#6 RWT620-6 'Profile & Front' #2
Test Record	Stamp: 10 Jan 1928	RWT620-11 EC #11 #2
Hand-written: 'Baird Phonovision record. Shows man's head in motion'	'Made in 1928'	RWT620-11 EC #11 #2
Hand-written: 'Baird Phonovision record. Shows lady moving head and smoking cigarette'. Autographed, J. L. Baird	'Made 28 March 1928'	RWT115-3 'Miss Pounsford'

Table 6-1. Summary data on all known Phonovision discs.

The reference numbers are consistent with the convention used by the Columbia Graphophone Company. Indeed three of the discs had Columbia 'Test Record' labels (see Figures 4-3 and 6-1).

Today almost no one has heard of the Columbia Graphophone Company. In its day it was world famous, encompassing household names in entertainment, such as Pathe, Odeon, Parlophone and of course Columbia.

The Columbia Graphophone Company

The name Columbia comes from the sales region of the original company, the Columbia Phonograph Company, formed in 1889 to license the sale of 'graphophones' in the Washington D.C. (District of Columbia) area.[1] In 1898, Columbia set up a branch in Great Britain. In 1917, the Columbia Graphophone Company was formed in Britain as an offshoot of its United States parent company. Two years later and the US parent suffered business difficulties requir-

Fig 6-1. Phonovision disc RWT620-4.
Courtesy of the NMPFT

ing re-structuring. It emerged under a different name, but continued to struggle, going into receivership in 1922. The British company, still trading as the Columbia Graphophone Company Ltd, became independent of its now defunct US parent. It became a public limited company in 1923 and went from strength to strength, eventually turning round and buying out what used to be its US parent, now called Columbia Phonograph Company Inc. In Europe, Columbia made further acquisitions, gaining the Parlophone (eventually to be the Beatles' label) and the Odeon label. In 1928, it acquired the recording business of Pathe in France making it the largest record company in France. With factories across Europe, in North and South America, Australia and Japan, Columbia became a world-famous name. In the United States, an investment by Columbia in broadcasting resulted in the start of the major network giant CBS, the Columbia Broadcasting System.[2]

In 1924, Isaac Schoenberg was recruited into Columbia and very effectively created the roots of what was to become one of Britain's foremost research and development laboratories. He recruited the highest calibre engineers, amongst whom was Alan Blumlein, considered by many to be Britain's best electrical and electronic engineer.[3]

By 1929, the company had reached a high point in business performance. Matters changed in 1930. The Depression had set in and the public were spending their dwindling resources on essentials at the expense of luxury items. Sales of records and domestic entertainment goods were hit hard. Under considerable threat, the Columbia Graphophone Company merged with its rival, the Gramophone Company (with its HMV – His Master's Voice – label). The merger in 1931 formed Electrical and Musical Industries Ltd (EMI). The EMI Group became one of the mainstays of the British entertainment industry until its decline in the 1970s led to the eventual merger with Thorn in autumn, 1979.

Dating 'Miss Pounsford'

Over half a century before, in the hey-day of the Columbia Graphophone Company, John Logie Baird had apparently engaged the Company's services in making test recordings, or at least test pressings of a recorded

= "RWT115-3"

= "Pounsford"

from existing RWT115-3
28th March 1928

Fig 6-2. The current 'Miss Pounsford' disc (RWT115-3) was found to have identical markings to a photograph that appeared in many publications throughout 1928. This is the evidence that these discs were made no later than 1928.

Courtesy of the Author

vision signal. One of these discs, dated 28th March 1928 and marked with a reference number and 'Miss Pounsford' on the disc surface, displayed a pattern on the recorded surface that looked like spokes of a wheel radiating out from the centre. It would be tempting to refer to this as a *classic* Phonovision pattern as it matched the official Baird Company photograph of a Phonovision disc. This photograph first appeared in the 'Television' magazine of July 1928 and subsequently in books.[4] It was wise to resist calling it classic, for when I studied the original 1928 photograph, it was very obvious that here was the same 'Miss Pounsford' disc. Considering so few Phonovision discs survived, it was amazing that one of them happened to have its picture taken as a publicity shot. This was of course not *exactly* the same disc; it was another pressing sporting a Columbia 'Test Record' label. However, the markings on the disc surface – including the hand-written 'Miss Pounsford' – were unmistakably identical (see Figure 6-2).

As this was the latest of all the surviving Phonovision recordings, this effectively confirms that all of them were made no later than mid-1928. With John Logie Baird's signature on the label, we can safely assume that this was genuine and not a contemporary fake. This assumption is safer than we might think, since Baird was the first (and seemingly the only) pioneer to attempt video recording.

Why are these Discs Phonovision?

Several other discs have similarities that classify them as being distinct from any other audio or video recording. Each is a single sided pressing that contains a 30-line vision signal. What makes these discs Phonovision is that each disc shows some attempt to have embedded timing – just like Baird's video recording patent. It was in that patent that Baird coined the term Phonovision to describe the process of synchronously recording the discs with the mechanical Nipkow disc camera. The process would ensure that the camera was mechanically geared to the record turntable so that a fixed number of turns of the Nipkow disc camera would result in one turn of the record turntable. The discs today all feature exactly 90 lines or three television pictures or frames per revolution.

If we find a disc that shows evidence of synchronous recording, it could be considered as a candidate for being Phonovision. The later recordings of the 1930s – the ones made by amateurs and the commercial test recordings – are in no way synchronised. The amateur discs are direct 'off-air' recordings of the broadcast vision signal. The commercial test discs show no evidence of being either directly attributable to Baird or to the BBC's television broadcasts. They appear to be part of a private venture by F. Plew, General Manager of the Major Radiovision Co. Ltd.

Valid Recording Dates?

The Phonovision discs are unique in that they have reference numbers on the shellac disc surface. In addition, most of them have dates stamped or written on the disc labels. Without corroborating evidence from company archives, we simply do not know whether the dates refer to the recording, the entry into the ledger or when the pressing was made. We would expect the reference numbers to be easily traced through company ledgers or logbooks. Unfortunately, no written record has survived. We can now only guess at what they meant.

When we take the dates along with the reference numbers, it looks like there were three recording sessions represented in the material we have today – in September 1927, January and March 1928. There could well have been other sessions, but the surviving discs are our only clues that there were such sessions. Two of the RWT620-s (see Table 6-1) were recorded on the 10th January 1928. The suffix on the reference number now looks like the number of the *take*. The fourth take survives from the September 1927 recording session (SWT515-4), three from the January 1928 session (RWT620-4, -6 and -11)))))and the third only from the March 1928 session (RWT115-3).

Baird's Recording Studio?

Commercial security was one of Baird's biggest concerns. His hard-earned television system was something that he believed he had a monopoly on, at least in the early days in the UK. He was not about to share the details of how he had achieved a practical television system with anyone, as this was his 'bread-and-butter'. Nevertheless he had to promote his work and his achievements. Though his achievements could be demonstrated publicly, he described his work and the progress he was making through articles and lectures.

Bill (W.C.) Fox had believed in Baird and his television invention from the very first successful demonstration of transmitting shadowgraphs down at Baird's lodgings in Hastings. Fox had been a news journalist working for the Press Association. He had been covering such widely differing topics as the Great War from a base on Northern India and the first run of the Northern Line Underground extension from its previous terminus at Golders Green to the then outlandish London suburb of Edgware.[5] He had supported Baird when he was starting out in television at Hastings and then at the time of the first demonstration of television to the members of the Royal Institution in January 1926. There he had shepherded those members, six at a time, into the demonstration room in Frith Street, London.

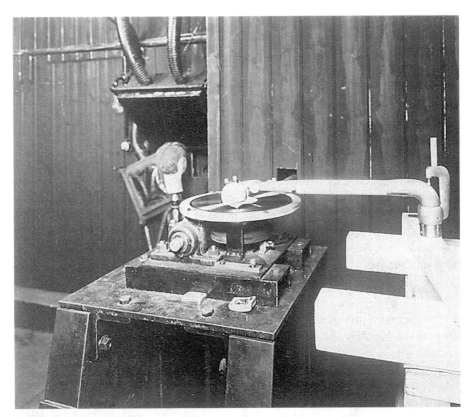

Fig 6-3. 1928 photograph of Phonovision equipment.
Courtesy of the Royal Television Society RTS 36-60

In 1928, Fox joined the Baird Company as their first and only Information Officer in charge of Marketing. Amongst other activities, he controlled the release of information and material to the press and arranged for photographs of the laboratories and of the equipment. These would be used to illustrate the progress being made. Amongst the many photographs that Fox released, there are only a few that are specific to what may be a Phonovision development laboratory. Their first appearance in the 'Television' magazine of July 1928 dates them to that time. Unlike some of the other laboratory photographs, no one knew exactly in which particular building these pictures of Phonovision were taken or, more surprisingly, what we were seeing in the pictures. The only surviving members of Baird staff that would have known in the early 1980s, Clapp and Fox, had no recollection.[6] Fox even featured in one of the photographs, however he admitted he was probably simply asked to stand in a certain place for the camera rather than being in any way involved.

At first glance, the photographs of the laboratory show a rather dull arrangement of hardware within the standard bare wooden panelling (sometimes called match-boarding or tongue-in-groove) that adorned the inside of Baird's laboratories.

On one photograph (see Figure 6-3), there is a record turntable, a long pick-up arm, a gear mechanism and a rod that is somehow connected with the turntable disappearing into a square hole in the boarding. Nearby a 'Stookie Bill' dummy head on a four-legged stand faces a large rectangular hole in the wall. Either side of this opening and above it, large black-covered boxes are being fed with wide flexible conduits. If this were indeed a television recording studio, it is a strange scene. There is no evidence anywhere of any electronics, wires, electric motors or accumulator batteries (the main source of power in those days). The equipment used for cutting discs is absent. This may well have been discarded or in use elsewhere, or even removed for commercial security reasons. It is also possible that, if this equipment was used for the Phonovision discs, Columbia may have made the recordings and then taken their equipment away after the experiments.

A second photograph (see Figure 6-4) shows a detail of the record turntable from the side. It now becomes a little clearer how this works. The turntable is a custom built device mounted on a helical gear arrangement. There is no motor on the turntable; the power comes from the rotation of the horizontal rod. Off to the left of the picture, somewhere behind the

Fig 6-4. Detail of the Phonovision deck.

Courtesy of the Author

panelling is a motor connected to the rod, which acts as a drive shaft. As there is a universal joint just before the turntable, the drive shaft is probably connected to the motor, either through further gearing or possibly as an extension of the motor spindle. The universal joint allows for physical misalignment of the turntable assembly with the motor. At the turntable the drive shaft is held by two heavy-duty bearings, between which is the helical gear. This meshes with another gear wheel mounted underneath the record turntable.

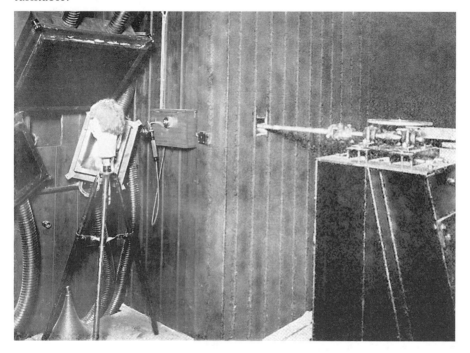

Fig 6-5. Another view of the Phonovision equipment on the right showing Noctovision equipment on the left. The dummy head faces three infra-red shielded lights and a large aperture through which parts of two large lenses can be seen.

Courtesy of the Author

No single picture shows all the detail for the deck and the gearing, but taken together we can understand how it would have worked. This gear ratio, the ratio of turns of drive shaft to turns of the turntable, is simply the ratio of the diameters of the two gear parts; the one on the drive shaft and the one underneath the turntable. We can see one of the gears in each picture. We cannot measure the exact dimension, but we can relate the diameter of each gear to that of the turntable, which we can see in both pictures. Doing this gives us a ratio of 3 to 1; the turntable turns once for every three turns of the drive shaft. We would expect any discs recorded on this equipment to have some multiple of three TV frames per revolution.

With the disc turning conventionally clockwise looking down on the turntable, we can follow this through the gears to see that the drive shaft was rotating clockwise as we face the wall.

Baird's Phonovision discs have one major feature that distinguishes them from any other disc recording of video. They have exactly three television frames on each revolution. Suddenly there is the high likelihood of a match between the Phonovision discs and the pictures of this laboratory. Indeed the date range of the discs – September 1927 to March 1928 – is consistent with the picture of the laboratory, which must have been taken (from its appearance in print) no later than mid-1928.

If this equipment was designed for the existing Phonovision discs, it would mean that, through the helical gears, the drive shaft would be turning at one revolution per picture, the same speed as the Nipkow disc, in a clockwise direction into the wall. It looks very much like there was no further gearing, and that the drive shaft was directly connected to the Nipkow scanning disc and drive motor.

All at once, the dummy head off to the side facing the large vertical hole in the wall takes on a different perspective. There is another picture of this dummy head in front of the blacked-out light-boxes. The caption for the photograph describes this as *Noctovision*. The light boxes have been covered with black ebonite sheets that block visible light, letting through near-infra-red light. The pipes leading in and out of the boxes allow air to flow over the light bulbs to stop them overheating and burning out. However, this picture is a selective enlargement from an original picture (see Figure 6-5) that covers a much wider area.[7] It reveals yet another view of the Phonovision equipment. The detail from the Noctovision picture of the hole in the wall shows parts of two circular objects – lenses mounted in a flat, possibly wooden, surface behind the hole.

The Studio in 3D

It was at this point that I took all three pictures, correlated the same objects in each view and extracted the dimensions into a computer model for what may have been the recording studio for Phonovision.

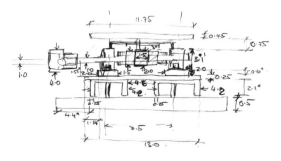

Fig 6-6. Original sketch of dimensions extracted from Figures 6-3, 6-4 and 6-5.

Courtesy of the Author

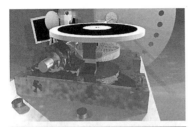

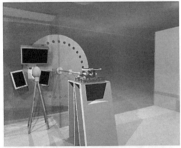

The absolute dimensions came from using the diameter of a surviving Phonovision disc, to scale with the one on the turntable, and the standard 4½ inches (11.43 cm) width of the vertical panels of match-boarding. The approach used nothing more sophisticated than a ruler measuring the horizontal positions of the objects in the room, helped, after some struggle, with my recollections of school geometry (see Figure 6-6). All the objects in the scene – the walls, the lights, the equipment and the dummy head – were reduced to a list of simple three-dimensional shapes and entered into a software application for rendering into a simulation of the scene (see Figure 6-7).

The computer-based visualisation of the room drew attention to the large vertical hole in the wall in front of the dummy head. The drive shaft turned clockwise once per picture as it entered the panelling and was probably connected directly to the Nipkow disc camera (see Figure 6-8).

The detailed picture of the dummy head facing the hole in the wall shows part of two large diameter lenses behind the hole. Their vertical spacing and the large size for the hole suggest a large disc behind the panelling. In fact from the 3-D model, the vertical spacing of the lenses is perfectly consistent with a disc some 1.5 metres in diameter centred on the drive shaft. With the drive shaft rotating clockwise, the lenses in the large Nipkow disc would sweep vertically from bottom to top in the aperture in front of the

Fig 6-7. Computer-simulated views of the laboratory.

Courtesy of the Author

dummy head on its stand. This is the scanning direction both on the Phonovision discs and in Baird's 30-line format (see Figures 6-8 and 6-9).

The space between the large Nipkow disc and the drive shaft entering the panelling looks to be big enough to house the electric motor powering the system.

The size of the aperture hole in the wall is several centimetres too large to mask off the area scanned by the camera. Of course, masking to the left and right should not really matter, as the bulk of the solid scanning disc should block any stray light. The simulation shows why the hole is on the large side. If the hole exactly matched the scanned area, the masking effect of the hole would reduce the light falling on the lenses. The lenses for the first and last lines would have half of their area blocked off giving fading at the left and right edges of the scene. The hole has been made wide enough to avoid that effect.

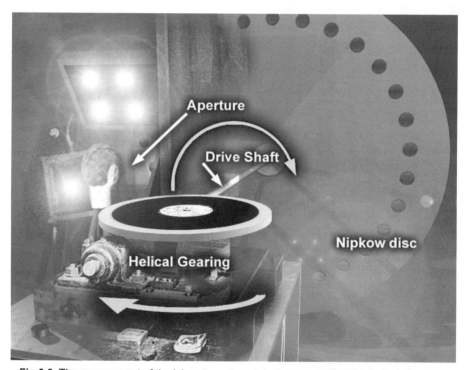

Fig 6-8. The arrangement of the laboratory strongly indicates a large Nipkow lens disc of around 1.5 m (5 feet) in diameter. From the direction of rotation of the turntable, following through the gearing, the drive shaft turns in a manner consistent with the Nipkow disc.

Courtesy of the Author

Fig 6-9. Computer simulations of the Phonovision deck (top) and laboratory (below).
Courtesy of the Author

When were the Photographs taken?

The large amount of dust or swarf on the close-up of the record turntable, and the scratches and dents on the paintwork of its main supports suggest that the equipment had been built and used for some time when the photographs were taken. Considering how sensitive Baird was to releasing information on how his systems worked, seeing these pictures comes as a bit of a surprise. In all of his published pictures, we rarely see the inner workings of his apparatus. It looks like the pictures were taken at the completion of Baird's work on Phonovision or at least Phonovision by this method.

We know that Baird moved from Motograph House to Long Acre in late 1927. If the photographs were taken in mid-1928, they must have been taken at Long Acre. If so, the equipment shown in the photographs appears to have been in use for both Phonovision and Noctovision within that period.

'Phoney' Phonovisor

Amongst the few photographs we have of the Phonovision experiments, there is one of J. D. Percy and W. C. Fox either side of equipment related to Phonovision. The caption to the picture in the book by Moseley and Barton Chapple reads 'A stage in the Phonovision Process showing some of the Early Experimental Apparatus'.[8] I had a long debate with Tony Bridgewater over this photograph, as it does not appear quite right (see Figure 6-10).

To my eyes, this appeared to be a prototype of the Phonovisor. The patent for the Phonovisor used a Nipkow disc mounted underneath and coaxially with the record turntable. For Phonovision there would have been ninety apertures in three spirals around the edge of the Nipkow disc. With a neon lamp placed underneath the edge of the Nipkow disc, the light would shine through the apertures and we would see the replayed image. Bridgewater agreed that this looked like a Phonovisor set-up but he had a problem with the 'neon'.

> 'To me the lamp exactly resembles a projection lamp of a type commonly in use in 1928 onwards for spotlight scanners. This had a bunched-filament and was usually 1000 watts. Note the heat-resistant white ceramic holder.'[9]

Bridgewater therefore viewed this as being a set-up for *recording* images. However there is a problem with the lamp not being shrouded. In every instance of spotlight scanning, the projector lamp sat in a blacked-out box similar to a *magic lantern*. The light from the unshielded lamp would

Fig 6-10. Analysis of the Phonovisor picture (see text).
Courtesy of the Author from original courtesy of Royal Television Society RTS36-23

ruin any image coming back from the scanned area. Whether this depicted recording or playing back is a minor point since the equipment could indeed be used for either – a video recorder of sorts. In either instance, this equipment could only have been used in total darkness – there was no viewing tunnel, no shrouding for either a lamp or a neon bulb. The main point is that the picture looks false.

Bridgewater also viewed the positioning of Fox as a 'contrivance'. Probably for photographic reasons, Fox was positioned on the wrong side to either see the replayed image, or to be recorded. However, the problems become obvious from analysing the detail of the equipment from the original plate.

The Phonovisor-like equipment uses *exactly the same chassis* as in the shots of the studio, right down to the dents and scratches in the paintwork. In the shadow, enhanced by computer, we can see the same helical gear assembly meaning this equipment was intended to be driven by a horizontal drive shaft. Off to the left, we see the universal joint of the drive shaft, disconnected and hanging down.

Studying the inverted-U mounts, we can see that the bolts that secure the chassis to the metal table in Figure 6-4 are missing in Figure 6-10. It appears that the turntable mechanism is not secured to the table.

With high resolution scanning to the limit of the photographic process, we can distort the picture to make it appear as if we were looking down on the Nipkow disc (see Figure 6-10 middle). Strangely, the apertures do not show up at all. This appears to be a plain solid disc.

Although we might possibly be seeing 'work in progress', it is more likely that, for all the reasons above, this photograph does not portray a real development at the Baird labs. Though this may appear shocking, it appears to have been common practice, as Bridgewater recalled in 1985.

> 'One of the problems common to many pictures of Long Acre activities is the fact that photographs for the press were often less than strictly authentic. Sometimes so as not to give away "vital secrets" – other times simply to move apparatus around to facilitate the photography in, as likely, a small room.'[10]

Radial Tracking

When we look at the photographs of the Phonovision laboratory, the long pick-up arm looks out of place. There are no wires connected to it and the arm has been placed on the deck with the needle angled to dig in to the disc rather than to trail across it. This looks incongruous and possibly staged for the sake of the photograph. However there is a good reason for having such a long pick-up arm.

From the almost perfect radial structure of the 'Miss Pounsford' disc, we can tell that the Phonovision recording equipment used a radial tracking cutter and that there was synchronisation with the Nipkow disc. In the heyday of vinyl discs, a radial, or more commonly called a tangential tracking record deck was what every audio enthusiast sought. Needle-in-the-groove discs were recorded in this manner all the way back to the days of 78 rpm shellac. Such a cutter would be arranged on a gantry centred on the spindle and would track in towards the spindle along a radius. Whereas for audio, playing back on a conventional swing-arm pick-up creates only slight distortion (mis-tracking), for the Phonovision discs, it would ruin any special synchronisation. To get a stable image on playback we would need to mirror exactly the recording method and use a tangential-tracking record deck.

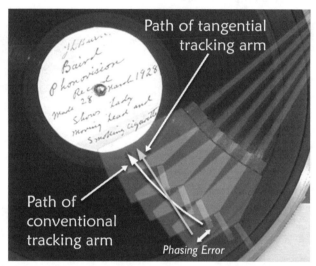

Fig 6-11. Multiple exposure illustrating the phasing error caused by playing back a Phonovision disc (such as the 'Miss Pounsford' shown here) with a radial arm.
Courtesy of the Author

A conventional playback arm (such as shown in Baird's patents and on the Phonovisor mock-ups) moves in an arc across the disc in a manner quite unlike the parallel tracking (see Figure 6-11). Putting some numbers on it, with a perfectly recorded disc and a 25 cm long pick-up arm, the image would appear to roll (vertically) roughly twice in the course of the complete disc playback.

We can see from the photographs (Figures 6-3 and 6-4) that the arm has been specially extended. The long radius of swing of the pick-up arm would give a correspondingly more stable playback. However, even with such a long pick-up arm, there would still be a slow vertical roll in the framing of the picture of about half a picture in height. Another possibility is that the extension was used to clear a large Nipkow disc, such as for the Phonovisor experiments (Figure 6-10).

Pick-up Type

The first electrical pick-up for sale was the 'Panotrope', appearing in December 1926. Several other types followed this and by late 1927 there were many to choose from. One of the better devices (and at £4, more expensive than the Woodroffe/Glasscoe used by the BBC) was the S. G. Brown pick-up. This is the device fitted at the end of the long pick-up arm in Figures 6-3 and 6-4 and shown in detail on Figure 6-12. The S. G. Brown as with other pick-ups received critical review, with evidence of multiple resonant frequencies on playback.[11]

Fig 6-12. The electrical pick-up shown in the Phonovision laboratory in Figure 6-4 (left) appears identical to the S. G. Brown pick-up (right) of 1927.
Courtesy of the Author (left), from original in Gramophone magazine (right)

The Bar reveals all

Simply marked 'Television' the first video recording is dated 20th September 1927. The disc and its label hold no other clue to its content. It is so heavily distorted by rapid variations in recording speed that the image could never have been seen clearly before computer processing.

Very fortunately, and quite uniquely, the start of this recording features what appears to be a horizontal bar or rod that looks white against the black background. Simply having this white bar in the recording gives us the material to find out a great deal about this recording, or at least the faults in it. It provides us with a reference against which we can take absolute measurements of the faults. The most serious of these faults affected the speed of recording and was the biggest impediment to viewing this disc.

Timing Faults

Right throughout the disc, the image rolls violently in an apparently random manner. The speed changes by plus or minus 2% quite rapidly, even within the time for one TV frame. This does not sound very much, but half that amount can turn what was a stable picture into one that immediately starts rolling vertically once every two pictures (see Figure 6-13).

The bar is horizontal, going across all lines in the TV frame as the lines are scanned vertically upwards. Though the bar appears to wiggle rapidly and randomly, there is no step change in its position from the last line of one TV frame to the first line of the next frame. This means that the bar is steady and perfectly horizontal, maintaining position throughout the whole time it is visible, at the beginning of the video recording. It is most likely to be a painted white line on a black card as, when it eventually is removed,

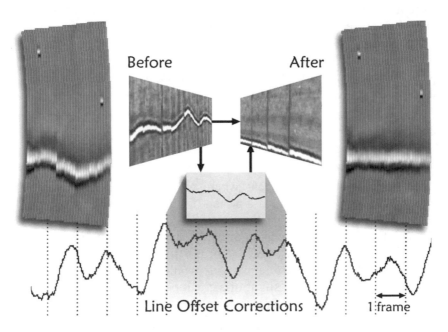

Fig 6-13. The uncorrected sequence of frames from the start of the September 1927 recording is passed through a line correlation process. This determines how much to move each line to create a stable image. A list of Offset Corrections, one per line, is saved for subsequent analysis of the cause of the fault. The graph shows 14 frames of which the central 4 relate to the 4-frame image sequence illustrated. Applying the corrections gives the stable sequence of images on the right.

Courtesy of the Author

we see the dummy head on its stand. At the time, this would have made rather a useful test-card for establishing any errors in fitting the lenses on the Nipkow disc camera. Now, it provides an extremely helpful reference to study the timing fault.

The owner of the disc, Ben Clapp, suggested that the disc might have been slipping on the turntable. Bill Fox subsequently told me a remarkably similar story. However, for that type of fault, we would expect to see more rapid, larger and even more catastrophic changes in speed than we do. Could there be some other reason for this effect?

The first step in restoring this recording was to convert the wriggling writhing white bar back to its original solid horizontal appearance. The sheer magnitude of the fault would suggest that this might not be totally random. We do not see anything like this speed variation in audio recordings made at the time. There had to be a source for this fault and probably one associated with the attempt to synchronise the recording with the Nipkow disc camera.

The white bar lends itself to simple line-by-line correction. This would restore the bar to being stable and horizontal. If we save the values for the line-by-line corrections as a sequence of numbers, we can use this to study what is causing the problem. These corrections represent how much the timing has departed from the original perfectly regular sequence. The trend represents the fluctuation in speed as variations in line-length, from one line to another. Just studying the plot of the trend shows no obvious pattern.

We need to look at the numbers differently, and an excellent tool for doing so is Fourier analysis. The results of the analysis, plotted on Figure 6-14, show the presence of several frequencies in this trend, most of which we expect but at least one that we do not.

The peaks in the graph occur where we have a periodic signal. The lowest frequency peak occurs at a period of 90 TV lines, one revolution of the turntable. This means that the signal is going first faster then slower with each turn of the Phonovision disc. We would expect to see some sort of peak here in any case just because of the difficulty in centring the disc. On playback, the pick-up arm and stylus would weave radially in and out once per revolution. Even a slight offset can be sufficient to cause a slight 'wow' and give rise to such a speed variation.

Not surprisingly the analysis also shows a variation in speed every 30 lines – at the TV picture frequency. There are at least two reasons for this. First, there is a periodic raggedness to the picture that repeats every frame. We will investigate this later in the chapter. Second, the bar was marginally off horizontal. This second effect also may account for the higher even harmonics at 2 and 4 times the TV frame rate.

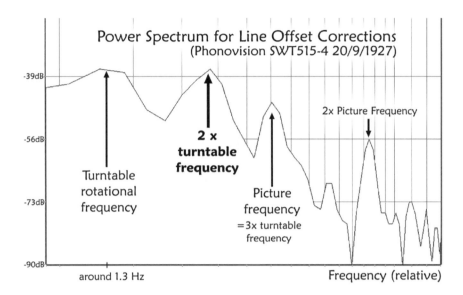

Fig 6-14. The power spectrum of the line-by-line corrections showing the presence of several frequency components, strongly suggesting that the instability is caused by a mechanical resonance in the system. The anomalous frequency is at the 2nd harmonic of the turntable rotational frequency.

Courtesy of the Author

All these frequencies are what we might expect, but there is one major anomaly. There is an unusually large component at twice the rotational speed of the Phonovision disc – the 2nd harmonic of the turntable rotational speed. Could playing back this disc slightly off-centre cause this?

The answer is simply no. Off-centre playback would give components at even harmonics, but working out the effect shows that the 2nd harmonic should be around 1/200th of the size of the peak at the turntable rotation rate – the 'fundamental' – for positioning the disc off-centre by as much as 2 mm. The peak in the graph at this 2nd harmonic is *equal in size to* the fundamental and the numbers show it is out of phase with it. There is some other cause, possibly from the mechanical layout of the recording turntable, though we really need some knowledge of the studio layout to understand what could be the source of the problem.

Classical mechanical engineering gives us two instances that would give rise to exactly this effect. In both instances, the effect would occur in a mechanical coupling: either it is a bent coupling or there is something like a universal joint in the drive chain.

'Stookie Bill'

Once the white bar is moved out of the way, we see a rather grotesque face.

Throughout the rest of the recording there is no change of expression on this face. What an absolute delight! After decades of reading about Baird's ventriloquist's dummy heads – all called 'Stookie Bill' – we now have a video recording of one of them (see Figure 6-15 top and middle). Baird had used these in his experiments as an uncomplaining test subject for his television system. The word 'Stookie' probably comes from a corruption of stucco, a fine-grained plaster from which the dummy heads are partly made. However, its use in Scots relates also to being like a figure made of stucco, that is, immobile like a statue.[12]

Image Shape

After someone's hand passes over the face (see Figure 6-15 middle), we see that the same person is holding the back of the dummy head with the other hand and then rocks the head from side to side (see Figure 6-15 top). The small angle it makes at each end of the swing looks like the head is mounted on a floor stand at roughly the height of someone sitting in a seat. That height is consistent with the dummy head in the Noctovision layout. That the head and support can stand freely and also be rocked from side to side means that the stand is not a tripod but more than likely has four legs. Again, this supports the notion that we are looking at the laboratory Noctovision layout (see Figure 6-5).

As the head moves from side to side (see Figure 6-15 top), it shows quite clearly the effect of using a mechanically scanned camera. The face appears spread wide as it moves in the direction of frame scanning, and appears squashed into a narrow strip as it moves back against the scanning direction. There is of course no blurring at all in such a camera. Unlike its electronic offspring, a mechanically scanned camera system scans the aperture over the scene.

Wherever the aperture is pointing in the scene, the signal that is sent is the instantaneous light value from that point. As a result, the 'exposure' is extremely short (around a 10,000[th] of a second) and progresses through the scene in the time for scanning one TV frame.

This strange stretching and squashing was not the only unusual feature about the images. There was another, more subtle effect. I used to show these images on my computer's display in a vertical letterbox format with

Fig 6-15 (other page). (top & middle) The dummy head, one of Baird's 'Stookie Bills', from SWT515-4 dated 20th September 1927. At top, the head is rocked from side to side. The arm rocking it can be seen at extreme right. At middle, a hand sweeps over the face near the beginning of the recording. (bottom) 'Miss Pounsford' from RWT115-3 (28th March 1928) turning her head from one side to the other in just 12 TV frames. The natural speed of movement suggests that the picture repetition frequency had to be around just 4 pictures per second.

Courtesy of the Author

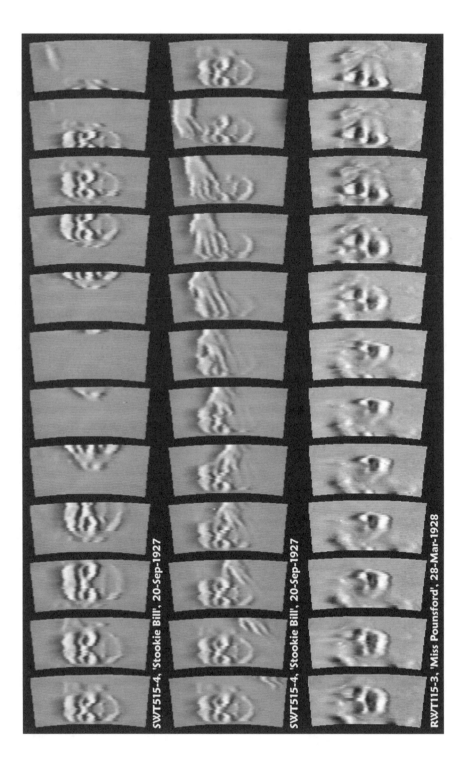

SWT515-4, 'Stookie Bill', 20-Sep-1927

SWT515-4, 'Stookie Bill', 20-Sep-1927

RWT115-3, 'Miss Pounsford', 28-Mar-1928

straight-line scanning. Shown like that, the head seemed slightly taller on one side of the frame, near the first line, than on the other, near the thirtieth line. Though puzzling at first, this effect was precisely what you would get if the camera was based on a Nipkow disc. Each line in an image scanned by Nipkow disc follows an arc of fixed angle rather than a line of fixed length. Each successive line is at a slightly different radius, so the physical length varies depending on whether the line is further away or nearer the centre of the disc. The image of the head is not actually taller; it merely occupies a greater fraction of the length of a line. It is the scanning line in the camera that varies in length rather than the dummy head.

On Figure 6-16, we can see, greatly exaggerated, the effect on the dummy head at either side of the image. Of all subjects to move from side to side, there could be none better than a 'Stookie Bill'. Being an inanimate object, the physical height of the head would not vary as it moved across each of the lines of the picture. I measured the proportion of the line length that the head occupied for each line that it passed across. The head rocked sideways several times, giving plenty of data points. Correcting for the rocking action, I plotted these values on a graph and extracted the 'best fit'. There was an excellent reason for doing this; the geometry of the Nipkow disc rather neatly and surprisingly allows us to measure the picture's aspect ratio.

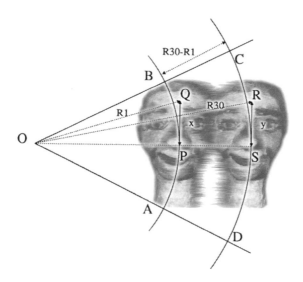

Fig 6-16. Diagram illustrating the effect seen on the Phonovision discs and measurable on the recording made in September 1927 of a 'Stookie Bill' dummy head.

Courtesy of the Author

Aspect Ratio

Aspect ratio is a term from both the cinema and television. It is simply the width of the image in relation to its height. The ratio 1.33:1 (that is, 4 units horizontal to 3 units vertical) used to be the standard for movies until the 1950s and for television from the 1930s to the start of the 21st century. Wide-screen cinema and television have a variety of aspect ratios with the most popular being from 1.85:1 to 2.35:1 for the cinema and 16:9 (or 1.78:1) for television. For his 30-line broadcast standard from 1929 onwards, Baird chose a 3:7 aspect ratio – a vertical letterbox.

In working out the aspect ratio for the 30-line Phonovision recordings, we only need to assume that the Nipkow disc had a single spiral of 30 apertures on each revolution. From the next 'fault', this is indeed the case. Purely from this, the gradient from the graph and some mathematics (see Annex) we get a value of 2.12:1 (vertical to horizontal). This is within 10% of the published broadcast format of 7:3 (that is, 2.33:1).

Faults in Building the Nipkow Disc

Masked somewhat by the rapid speed changes on the September 1927 disc, there is another more subtle fault. The picture has a ragged appearance where certain lines are slightly offset relative to each other. The pattern of this raggedness is the same in every single TV frame, throughout the recording. What we are seeing is that the start of each line in the TV frame is slightly different from its neighbours but consistent from one frame to the next (see Figure 6-17). That means if line 7, say, is early in one frame, it is equally early on *every* frame. The raggedness does not just appear on the September 1927 disc, it also appears throughout every take of the January 1928 session. It also appears on the March 1928 recording but is almost impossible to separate from that disc's own and unique timing fault. What caused the lines to have this fixed pattern?

If we think back to the formation of the picture in Baird's mechanical television system, each TV line has its own dedicated aperture. As we have just seen, the Phonovision discs were all scanned using a Nipkow disc camera. Each of the lenses in the Nipkow disc needs to be placed at precisely equal steps around the circumference to generate a stable picture. Any errors will show up as the lines appearing to start too early or too late. In reality it is the *image* cast by the lens that is out of place.

Stacking the 30 lines of each TV frame like a pack of cards seen edge on, we would get a perfect stack if we rested the pack on one end on top of a flat surface. If however, the surface was not flat but uneven, then the stack of cards would follow the surface, with some shifted up or down in a pattern that followed the unevenness of the surface.

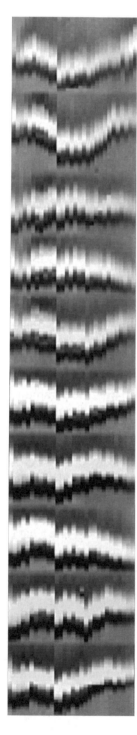

The errors vary in size with the biggest occurring between lines 11 and 12. There the error amounts to about 4% of the length of one line. This sounds quite a large error, but we need to understand that we are talking about the error in only one thirtieth of the circumference – the angular distance between lenses. For a disc 1.5 metres in diameter, such as we have deduced from the studio photographs, the 4% error would amount to placing one of the 10 cm lenses just 7 mm, one fourteenth of the diameter, slightly too far round the circumference. Baird's first assistant, Ben Clapp, recalled making Nipkow discs of around 5 feet (1.5 metres) in diameter.[13] When I told him of this error, he remarked that the discs were hand-made and experimental. That level of error was not at all surprising.

Like most artefacts to do with vision, this fault looks much worse than the numbers would suggest. A single static picture shows the fault, given the right image. The ragged pattern becomes much more obvious when we see a moving image, and particularly if an object moves sideways across the scan lines. That this pattern is consistent from one frame to the next throughout the recording means that the Nipkow disc camera had only one spiral of lenses rather than two or more.

The ragged appearance of the picture shows up the errors in positioning the lenses around the circumference of the Nipkow disc. On the Phonovision discs we cannot see any effect of errors in the other direction – along the radius. Such an error would easily show up on a Nipkow disc display. Holes placed too close together would cause extra brightness through overlap of the corresponding

Fig 6-17. The best illustration of the effect of misplaced lenses in the Nipkow disc camera comes from the white bar at the start of the September 1927 Phonovision disc. The picture on the right shows the bar from ten frames. Line 1 is on the left, 3 on the far right. The biggest step is between lines 11 and 12. Other lesser steps are apparent. (Each frame shows variations in trend caused by speed fluctuation, meaning we can only judge the disc fault from examining adjacent lines.)

Courtesy of the Author

lines and would squash the picture in width, whereas holes too far apart would stretch the picture and leave dark gaps between lines. The effects of circumferential and radial errors in the Nipkow disc were described in 1930 as the consequence of not taking care in making the disc.[14]

This effect and the ragged appearance shows up on the historic first off-screen picture of television in 1926 taken by the photographer Lafayette. There are very few off-screen images taken in those early days, probably due to the long exposures necessary to capture images from the faint displays of the time. When we view these images today, the Nipkow disc faults are obvious. The distortion in the photograph of Oliver Hutchinson, identified as the earliest picture of television, is completely caused by Nipkow disc faults (see Figure 3-13). On poorer quality reproductions, some historians have had difficulty interpreting what they saw, thinking that it may have been, say, a split image. As well as a gradual 'tilt' when going from left to right, the right hand side is shifted upwards relative to the left. We can see large gaps between the lines on the right. The Nipkow display disc simply has the outer eleven holes at too high a step in radius between them. It was almost as if the disc-maker took a tea break at that point.

The top of the forehead is uneven and the pattern of this unevenness is identical in the light region at the bottom of the picture. These patterns form what I view as a 'signature' for both camera and display scanning discs as they incorporate the combined errors on lens and hole positions around the circumference with that of the radial positioning. Comparing the famous picture of Hutchinson with another Lafayette picture (see Figure 7-9a and b) – both single-spiral 30-line images – shows the largest circumferential error occurring between lines 11 and 12, counting from the outside of the disc (assuming right-to-left picture scanning). As the vertical streaks of the display show no correlation, we can safely assume that these two pictures used the same camera but different display discs.

The positional errors on these photographs come from the combined effect of the Nipkow discs for both camera and display. It would therefore be wrong to try to compare what we see on Phonovision with the contemporary photographs. Phonovision only tells us about the camera. All we can say for certainty is that the errors are roughly the same value and that the camera disc used for Phonovision (1927 to 1928) would have created pictures no better than the Lafayette 1926 images.

Is this a surprise? Not really. The Phonovision discs pre-date the massive influx of professional engineers into Baird Television Ltd. The influx started at the beginning of 1928. The high quality of off-screen photographs that appeared in 1929 suggests a substantial improvement in

Nipkow build quality with no detectable errors. We should be a little cautious however.

There was a trick in getting off-screen photographs that was used to simulate 30-lines. Described in the 'Television' journal in 1931, this amounted to having the camera facing the Nipkow display disc behind which was a ground glass screen acting as a diffuser.[15] In front of that, mounted coaxially with the first disc was another, identical Nipkow disc on which a lens focused the scene. The result was of course excellent with perfect tonal quality and, of course, no evidence of phase errors. The technique was a way of demonstrating the principle of scanning. We should be suspicious of any off-screen picture from this period that shows no raggedness in lines and no effect of phase errors (see Figure 3-14).

Progress in Timing

All the timing faults so far relate to the earliest Phonovision disc. The later sessions – in January and March 1928 – show different timing faults but with signs of development and improvement. The errors we see from the misplaced lenses in the Nipkow disc also show up on the January 1928 recordings, telling us that the same Nipkow disc camera was in use.

The January 1928 recordings feature a man's head (see Figure 6-18). Although the head moves around quite a bit, there is almost no change of expression on the face. This was probably one of Baird's engineers rather

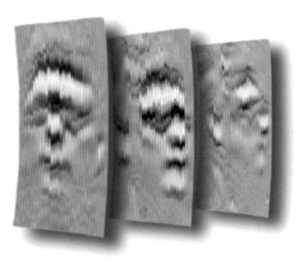

Fig 6-18. Three processed frames from a Phonovision recording (RWT620-11) dated 10th January 1928 of 'Wally' Fowlkes. Made over 3 months after the Recording of 'Stookie Bill', these more stable recordings are the earliest known video recordings of a person.

Courtesy of the Author

than an actor. If the name 'Wally' scratched on the disc surface is anything to go by, then this is Wally (Walter) Fowlkes, a junior technician-assistant working for Baird (see Figure 6-19). Wally had been with Baird since late 1926. At that time, his 'only assistant was the office boy imported from Hutchinson's soap works'.[16] Baird seems to have used Wally as an animated version of 'Stookie Bill'. When Baird was trying out other non-visible light sources such as ultra violet, far infrared (heat) and near infrared, Wally was Baird's prime viewing subject.

> 'He was ignorant but amiable. The ultra violet rays affected his eyes, and he did not complain, but I got a fright and tried infrared. I first used electric fires to get these infrared rays, which are practically heat rays. I could not get any result and added more fires until Wally was nearly roasted alive, then I put in a dummy's head and added more fires and the dummy's head went up in flames. I decided to try another track and use the shorter infrared waves. To get these I used ordinary electric light bulbs covered with thin ebonite. This cut off all light but allowed the infrared ray to pass. Wally sat under this without much discomfort and after one or two adjustments I saw him on the screen although he was in total darkness. That was again a thrill, something new and strange, I was actually seeing a person without light.'[17]

Baird's anecdotal description of his dabbling in what was to become Noctovision gives strong support to the man on the Phonovision discs being indeed Wally Fowlkes. With the laboratory photographs showing Phonovision apparatus right next to that for Noctovision, and 'Stookie Bill' and Wally as the prime test objects, it is not all that surprising that the two earliest surviving recording sessions for Phonovision show a 'Stookie Bill' and 'Wally'.

By the time of these recordings of Wally, Baird had improved the stability of his timing though there were still residual errors and long-term drift. However, these discs had nothing like the white bar on the 'Stookie Bill' recording, making it difficult to measure the improvement.

Fig 6-19. Wally Fowlkes, (left) from a group photograph taken at Long Acre in 1929 and (right) from the roof of Long Acre 1928.
Courtesy of R. M. Herbert (left)
From Larner, 'Practical Television' (right)

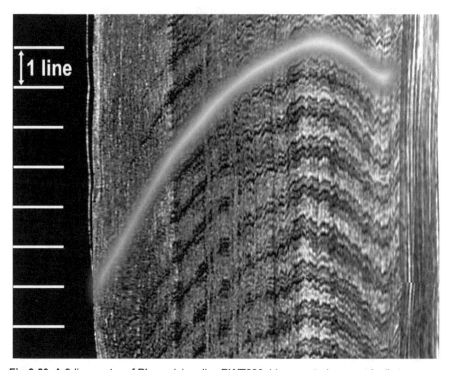

Fig 6-20. A 9-line sector of Phonovision disc RWT620-11 converted geometrically to show the drift in line position throughout the recording

Courtesy of the Author

As Phonovision was designed to have a precise number of frames per revolution, the side effect is that the video on each line of every third frame aligns up radially. The perfect radial structure of the March 1928 recording shows what happens when this works properly. The January 1928 recordings do not show such a perfect pattern. On top of the residual timing errors, there is gentle drift in speed throughout the discs. On the last take, RWT620-11), this amounts to the turntable gaining on the Nipkow disc by about 6 line-lengths in around 700 frames, or 0.03% slightly faster for most of the recording (see Figure 6-20). Though the perfect radial *spokes* are not there, could these recordings really have been synchronised with the camera scanning disc?

Howling Evidence

The proof that they were came from a strange sound at the start of all these January 1928 recordings. This had puzzled me literally for years. It was not until I plotted the sound as a sonogram that the answer appeared. At the very start of all the recordings from the January 1928 session, there is a sound that drops rapidly in pitch and stabilises. On one disc there is an accompanying whistle that drops in pitch becoming a loud howl after a few seconds. Along with that sound we can hear the sound of the video waveform.

The sonogram is normally used to illustrate the component frequencies of speech. It shows frequency vertically, time horizontally and intensity of sound as shades of grey. On the disc with the whine and howl (RWT620-6)), we see the pitch falling down and shown as a black line – a single note dropping in frequency (see Figure 6-21). To our ears this whine or note fades out after a few seconds. We can see on the chart though that the sound continues through, becoming the strongest frequency in the howl on the right of the chart. This howl appears on all these recordings to varying extents with the whole of take 4 (RWT620-4)) almost completely ruined by it, take 6 (the one used here – RWT620-6)) partially affected by it and take 11 (RWT620-11)) nearly free of it. Noticeably triggered by the vision signal going above a certain amplitude, this howl is most likely Baird's video amplifier going unstable and bursting into oscillation.

This amplifier was used to boost the tiny signal from his photocell positioned behind the Nipkow disc and collected the light from each of the lenses as they swung by. In the 1920s, electronics and high-gain low-noise amplifiers in particular were state of the art. Even with large lenses to gather as much light as possible, there was never enough for his photocell. Consequently he flooded the area with light, used the biggest Nipkow disc

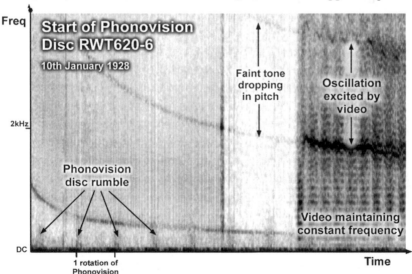

Fig 6-21. Sonogram of the start of a Phonovision disc from January 1928 (RWT620-6).
Courtesy of the Author

camera that he could build and used the maximum achievable gain on his amplifier.

Gain goes with pain in the analogue hardware and especially the video amplifier designer's world. The problem with high-gain amplifiers is that, beyond a certain gain, coupling of the amplified output back to the input causes a runaway feedback loop – much like a microphone on a public address system – leading to an electronic howl or oscillation.

The sonogram shows us that this note falling in pitch is an oscillation in the amplifier. However, it is very unusual to have the pitch dropping in frequency – especially since the other two discs from this session also have other sounds right at the start also falling in pitch. The graph of the fall in pitch, shown on the sonogram, gives the answer to this dilemma. In reality the oscillation is not falling in pitch, it is at the same pitch throughout and the recording turntable and disc is accelerating up to speed.

Taking a second look at the sonogram shows the video signal, or rather the frequency components of the video signal, appearing like Venetian blinds – a stack of horizontal bands for each of the frequency components. Unlike the oscillation falling in pitch, these bands are perfectly horizontal throughout.

These horizontal bands show that, whilst both record turntable and camera disc are accelerating up to speed, they are doing so exactly in step, which can only happen if they are mechanically connected. This is quite amazing – direct evidence that the record turntable is linked physically to the Nipkow disc camera for the January 1928 recordings.

Interestingly in the first few seconds of the discs, we hear a rumble dropping in pitch at the Nipkow disc rotational rate. This is the sound of the Nipkow disc coming up to speed, probably vibration coupled through the drive mechanism linked to the turntable.

Torque of the Drive Motor

Studying the profile from the sonogram, we can work out how fast the Nipkow disc came up to speed. The disc cutter started recording when the Nipkow disc had reached 20% of its final speed after just over two rotations. In the following six rotations, it had achieved 50% speed and after a further 24 rotations had reached 90% of its final speed. If we knew the mass of the Nipkow disc, with this acceleration profile we could estimate the torque force of the main drive motor.

Though the recordings were synchronous, an explanation is lacking for the drift in speed throughout the recording. The spokes that should be radial, form a herringbone pattern. This amounts to the turntable going

0.03% faster than the reduction gear should allow. For perfect hard-locked synchronisation, we would expect to see this pattern perfectly radial, like the spokes on the 'Miss Pounsford' disc. Such a gentle drift may come from having somewhere in the drive chain a clutch mechanism or some type of flexible link.

Miss Pounsford's Cigarette

When it comes to the March 1928 recording session, we only have one example, one 'take' to figure out what was going on. Unlike the recordings from the earlier sessions, this disc shows a straight-line spoke structure. A natural assumption is that the synchronisation was perfect. In 1983, the excitement of transcribing this unique and precious disc turned rapidly to horror when I tried to view it. This disc had perfectly registered TV frames, but there was a new and serious problem with the 'lady moving head and smoking cigarette'. 'Miss Pounsford' would elude me until I got down to writing more software.

There is something emotive about cigarettes and television, though not from the later point of view of commercial advertising. Historic descriptions of some of the early shadowgraph images describe the action of blowing smoke in silhouette. The billowing movement of the exhaled smoke appeared life-like. The action consequently played a key role in early demonstrations of television. As a result, when we come across the 'Miss Pounsford' disc with the label sporting 'Woman smoking a cigarette' we start getting excited.

When we play back the disc straight, without processing, we see a stable image with the vague impression of a head. We can occasionally see what looks like a cigarette hanging down from what might be a mouth. The disc has been demonstrated in this way many times, but the image seemed strange and surprisingly difficult to interpret (see Figure 6-22).

By whatever method Baird had used to achieve perfect synchronisation of TV frames with each other (hence the radial pattern on the disc surface), it had caused something else to suffer. Within each frame, it looked like the lines were stacked, just like the pack of cards analogy that explained the effect of misplaced holes. However, unlike the misplaced holes, the lines seemed to rest on a triangular or zigzag surface. There is of course no real surface on which the lines are stacked. We are seeing a pattern of errors in the start point for each line that makes successive lines arrive early or late. This is however not random, like the misplaced lenses that we saw earlier. Also, this does not remain the same pattern throughout the whole recording. For the first few seconds there is no error, but the error suddenly grows, then stabilises, remaining for the rest of the recording.

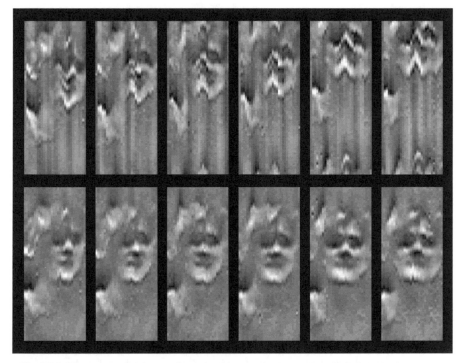

Fig 6-22. Six consecutive frames from the Phonovision recording of 'Miss Pounsford' dated 28th March 1928. The sequence at top is without timing correction and illustrates how the timing fault can create artefacts. The sequence below incorporates timing correction and removal of disc playback rumble.

Courtesy of the Author

There is a disappointment in store if we think we are going to see a cigarette and billowing smoke. 'Miss Pounsford' is not smoking and has no cigarette. The 'cigarette' hanging down from her mouth is, in fact, her chin, grossly distorted by the recording fault.

It is convenient to think of the new fault as being associated with rock-solid frame synchronisation. The earlier recordings neither have this fault, nor have evidence of such perfect and stable frame synchronisation. From the high frame-level stability, it looks like the 'Miss Pounsford' recording was the one and only recording to have a directly coupled mechanical linkage to the Nipkow disc. This direct coupling more than likely allowed a vibration to start up in the equipment. Because it appears on every frame rather than every third frame, the vibration is probably associated with the drive linkage from the Nipkow disc and motor and possibly any item in the drive chain – especially a universal joint.

Looking back, we can see a progression, a development, in the quality of the timing. 'Stookie Bill's image in September 1927 suffered rapid high speed changes caused by some problem with the drive mechanism for the turntable. This had been fixed by the January 1928 sessions with 'Wally' Fowlkes as a living subject. Though there was synchronisation, there was still a drift, probably through a flexible coupling or drive. Almost three months later and we have the non-smoking 'Miss Pounsford', synchronised but suffering from the effects of high speed mechanical vibration. We recognised 'Stookie Bill', we guessed at 'Wally' Fowlkes, but who was 'Miss Pounsford'?

'Miss Pounsford'

Once the timing distortion was corrected, the lady on the disc showed herself to be much less stilted and more natural than the man on the 'Wally' discs. In fact I had spent so intensive a time correcting the errors that I had not fully realised how impressive the recording was. When my wife, Lydia, saw the first results of processing, she was quite astonished by how much she could see. Being far more fashion conscious than me, she noticed the ringlets in the lady's hair and the shoulder straps of a low-slung dress. What took her by surprise was how outgoing and natural this lady appeared, seeming to blow a kiss, to laugh and talk quite animatedly. This 'Miss Pounsford' was quite a lady.

She was however a lady with a name that not one of the early Baird employees remembered. For years, I assumed that she was a visitor that simply happened to be in the right place at the right time. Of all the moving pictures from Phono-vision, 'Miss Pounsford' was the most popular with the media. On one occasion in 1993, the producer of a Channel 4 document-ary, 'The Long Summer' was keen to find out who she was and arranged for a plea for information on Miss Pounsford to go out with the images. Anne Cracknell just happened to be watching that prog-ramme and made contact with the producer. Miss Pounsford, Miss *Mabel*

Fig 6-23. A photograph of Mabel Pounsford (left) as she appeared in the 1940s, many years after her 1928 recording session on Phonovision disc (right).

Courtesy of Anne Cracknell & the Author

Pounsford, had been her great-aunt and godmother. She was born on 22nd June 1883 and had lived at Tulse Hill, near Streatham in South East London. Baird had employed her 'as a secretary, through an agency'.[18] The photocell Baird used was more sensitive at the red end of the spectrum. The fact that Mabel Pounsford was a redhead may explain why her hair showed up so well, whereas 'Wally' Fowlkes' on the earlier discs had not.

Scanning Direction

There was nothing in the video signal that could tell in which direction the picture or frame was being scanned – whether it was from left to right or right to left. I needed a unique subject that would look correct only one-way round. That unique subject came with the photographs of Mabel Pounsford provided by Mrs Cracknell. These suggested that the arc-scan was oriented like the contemporary off-screen snapshots of the 30-line picture, with the centre of arc off to the left (see Figure 6-23).

Though the aspect ratio and numbers of lines all comply with Baird's 30-line format, there is one feature that turns out different. The earlier measurement of the aspect ratio identified the first line in the TV frame innermost on the Nipkow disc camera and the thirtieth line outermost. With Mabel Pounsford's hair parting showing that the along-line scanning for the Phonovision image was counter-clockwise, this means that the frame had to be scanning from left to right – the opposite direction to the published 30-line Baird format.

In 1960, Geoffrey Parr recorded an introduction to what turned out to be the Royal Television Society's disc (RWT620-11)). He described the scanning direction as being from left to right. This is the opposite of Baird's 30-line format and may have been a simple slip of the tongue – were it not for the fact that we can now see that the Phonovision images were scanned from left to right.

Too Fast

Baird's 30-line television format with vertical scanning and 3:7 aspect ratio is mirrored on the Phonovision discs. However there is one part of the format that is difficult to establish and impossible to measure accurately. We cannot tell directly from replay what the recording speed was and consequently what the frame rate was. If only there had been some verbal identification or some unique sound that could give a hint of recording speed. The vision signal, the video amplifier oscillation and some rumble from the drive motor were the only detectable signals on the recordings.

Baird's format called for 12½ frames per second with each frame comprising 30 lines, each of which consequently would be running at 375 lines per second. The discs all have 3 frames, 90 lines, per revolution. To give the Baird broadcast format of 12½ frames per second, the Phonovision disc would have to have been recorded at one third of that, around 4 revolutions per second or 250 revs per minute (rpm). Considering that standard wax masters and shellac discs were all designed for 78 to 80 rpm, we seem to have a dilemma.

There are a few hints as to why the conventional rate of around 78 rpm should be used. The most important came from Eliot Levin, an expert in early recording technology. In the course of professionally transcribing them for museum archive, he studied the Phonovision discs looking for anything unusual.[19] He said that it was extremely unlikely that these discs were recorded at 250 rpm. They showed all the characteristics of being conventionally recorded at a speed ranging from around 78 rpm up to a maximum of 120 rpm.

The clinching argument for a lower rate appears throughout the videodisc of Mabel Pounsford. Her actions are simply too fast to be realistic when played back at the Baird broadcast rate of 12½ frames per second. At one point, she turns her head to the right and then to the left – all within eleven consecutive frames (see Figure 6-15 bottom). This would be less than one second at the Baird broadcast rate. The movement looks far more natural at 6 frames per second or less.

'What is Wrong with this Picture?'

If there is one thing outstanding about all these Phonovision discs, it is the atrocious quality of the video signal. We would of course expect some problems simply with recording video as audio. We came across one of these faults in the previous chapter when discussing the importance of preserving phase information for video. A commercial recording of 30-line test stills (the *Major Radiovision* disc) made in the mid-1930s, has quite excellent recorded video quality when compared with Phonovision. What could be going wrong?

The Phonovision recordings have 'soft' edges suggesting no high frequencies. The grey background and lack of wide area tonal variation make them look as if they have had most of their low frequencies removed. The resultant signal looks more like a modulated sine wave than a video signal. We can use the white bar at the beginning of the 'Stookie Bill' recording yet again to help us understand what is happening. The trick here is to compare the replayed white bar signal with a simulated white bar. Using Fourier Analysis on a complete line of data, we get the amplitude of the harmonics of TV line frequency for both actual and simulated signals. The ratio of those amplitudes – actual to simulated – tell us how much attenuation has occurred at each of the harmonics. When we plot these on a graph, we effectively get the frequency response of the whole Phonovision recording and replay chain. This encompasses the effects of camera aperture, photocell response, amplifier response, response of the link with the recording cutter, response of the wax to the cutter and the response of the replay device (see Figure 6-24).

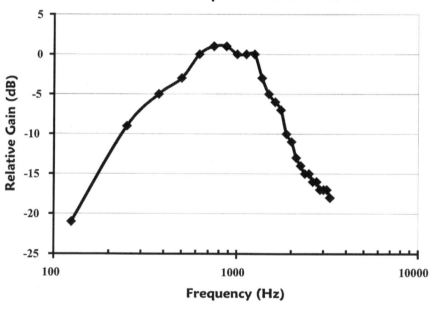

**Estimated Frequency Response
of Phonovision Recording**
SWT515-4 20th September 1927 - Test Bar

Fig 6-24. The estimated frequency response of the Phonovision system, from photocell to disc. This was derived from modelling the white bar test signal at the start of the 'Stookie Bill' dummy head recording (SWT515-4) and comparing its frequency characteristics with that of the unprocessed signal.

Courtesy of the Author

We know the frequency response of the disc cutters used by the Columbia Graphophone Company in the late 1920s and it is nowhere near as bad as what we are seeing on Phonovision. Suspicion falls on the link between the output of Baird's video amplifier and the groove in the wax. This whole area is unknown to us. It could be that some portable Columbia cutting equipment was adapted to Baird's synchronisation equipment. If so, then the poor frequency response must be caused by the link between the output of the video amplifier and the input to the Columbia disc cutting equipment. What we see is so poor in quality that the link may be no more than a loudspeaker in front of a microphone. Whilst such an arrangement might raise a few eyebrows, the Columbia engineers may have dictated that there should be no electrical connection with Baird's experimental equipment.

Another possibility is that Baird made his own wax masters and that the head-cutter was experimental. Suspicion would then fall on the disc cutting equipment. This is more likely as we have seen that in two of the three sessions, the record turntable was indeed mechanically linked to the Nipkow disc. Against this possibility is the high quality of the cut of the groove. Professional cutting equipment had to have been used.

Whatever the reason, the low video quality would in no way have helped Baird and his staff recognise the image on playback. Any understanding of the pictures would have been seriously hampered by all the differing timing faults, by the low picture refresh rate of around 4 frames per second and by the narrow range of frequencies that have been recorded on all these discs.

Work in Progress

That we see development in stabilising the image across the series of Phonovision discs, yet no improvement in the video quality, may mean that Baird was focused on overcoming the timing faults. These discs were after all only experimental and it could well be that once Baird had overcome the problems with timing, he would have tackled the poor video quality. Given that the recordings made a few years later on aluminium discs with Dictaphone quality sound were vastly superior in frequency content, this would have been as easy step for Baird to take.

Never intended for release to the public or for appraisal, these discs merely hold a record of work in progress, rather than any successful result. We should see them as offering us a unique peek over the shoulder of Baird at work, dealing with day-to-day problems in the attempt to develop a video recording and playback system in the late 1920s.

---

[1] ANON.: 'The Virtual Gramophone' (National Library of Canada), Internet, June 1998

[2] MARTLAND, P.: 'Since Records began: EMI the first 100 years' (Batsford), 1997, pp116–119

[3] BURNS, R. W.: 'The Life and Times of Alan Dower Blumlein' (Institution of Electrical Engineers), 2000

[4] DINSDALE, A.: 'Television' (Television Press), 2nd edn, London, 1928, facing page 140

[5] FOX, W. C.: Private communication, 1983

[6] FOX, W. C. & CLAPP, B.: Private Communication, 1983–84

[7] FLEMING, J. A.: 'The Inventor of the Fleming Valve visits the now World-famous Baird

Laboratory', *Television,* July 1928, pp5–7

[8] MOSELEY & BARTON CHAPPLE, 'Television Today and Tomorrow' (Pitman), 1931 facing p130

[9] BRIDGEWATER, T. H.: Private communication, 9th Mar 1985

[10] BRIDGEWATER, T. H.: *ibid*, 9th Mar 1985.

[11] 'Electric Pick-ups', *Gramophone*, Sep 1927, pp143–144

[12] 'The Concise Scots Dictionary' (Chambers), 1985

[13] CLAPP, B.: Private communication, 1983

[14] KNIPE, A.: 'How to Make an Accurate Scanning Disc', *Television*, Jan 1930, pp547–549

[15] CAMPBELL, D. R.: 'Scanned Images with Simple Apparatus', *Television*, Sep 1931

[16] BAIRD, J. L.: 'Sermons, Soap and Television' (Royal Television Society), 1988, p65

[17] BAIRD, J. L.: *ibid*

[18] CRACKNELL, A.: Private communication, 29th May 1993

[19] LEVIN, E. B.: *Symposium Records* for National Museum of Photography, Film and Television, Bradford, 1996

7   Television Develops

I met a traveller from an antique land
Who said: 'Two vast and trunkless legs of stone
Stand in the desert. Near them, on the sand,
Half sunk, a shattered visage lies, whose frown,
And wrinkled lip, and sneer of command,
Tell that its sculptor well those passions read
Which yet survive, stamped on these lifeless things,
The hand that mocked them and the heart that fed;
And on the pedestal these words appear:
"My name is Ozymandias, king of kings:
Look on my works, ye Mighty, and despair!"
Nothing beside remains. Round the decay
of that colossal wreck, boundless and bare
The lone and level sands stretch far away.'

'Ozymandias', Percy Bysse Shelley

The Legacy of Baird

In the late 1920s, John Logie Baird was by far the most prolific television pioneer. He explored every avenue of television's possibilities through implementing 'What-Ifs' and demonstrating the results, with maximum publicity. Whilst the shares in the Company rose dramatically as a result, Baird drove himself on (see Figure 7-1).

> 'I was not interested either in shares or money or stimulating progress, I felt I was doing something worth doing.'[1]

Those words capture simply what kept Baird focused. Like any creative person who has achieved success, he was being encouraged and carried forward by it. His 'never-been-done-before' demonstrations bred public interest and wonder. That encouraged funding to pay for more development work. The company had been a 'start-up' and needed the continued publicity to attract financial backing.

The demonstrations of television's possibilities can be thought of in two ways. They were either in the world-class category, ahead of all other pioneers, or they were mere stunts requiring little true research and development to garner media attention. When we appraise these demonstrations today, we see that they truly represent the *promise* of television rather than any innovations in the technology. They are of the 'Hey, look, you can do *this* with it!' category. We could attack Baird for undertaking what amounts to a series of stunts. But these stunts were all true 'firsts', capturing the public's imagination and reinforcing Baird as being the herald, though not necessarily the deliverer, of a new age.

Fig 7-1. Baird being televised in floodlit studio. This is a similar arrangement to that used in the Phonovision laboratory. The two horizontal wooden posts may have been used to mark out the limits of the camera's field of view. Stretched across the posts was a thin wire against which the subject being televised placed his or her forehead. This ensured the subject was at the right distance from the camera. (February 1928)
From original courtesy of the Royal Television Society RTS36-10

Over the years, some have considered these demonstrations as merely raising Baird's personal profile and creating a legend. That may be the wrong view to take. It is true that the demonstrations made Baird a contender for being the 'father of television', but there are other aspects to consider. So long as the demonstrations kept Baird as number one, the Baird Company was also number one. From the point of view of the financial backers, the Company must surely then have been in the best position to bring television to market and the best to capitalise on television.

We can get the issue of *stunts* into better perspective if we look at similar stunts done by the BBC with their 405-line service a quarter of a century later. The novelty of the BBC bringing live pictures from unusual locations fed public interest for a while. We don't think poorly of the BBC for those 'first television transmissions from...' broadcasts of the 1940s and 1950s and their 'firsts' as seen on their 'Saturday Night Out' series of live outside broadcast programmes. The BBC, of course, was not in it for the money – they were merely sharing their enthusiasm for the technology with the public. This time it was the BBC saying, 'Hey, look, you can do *this* with it!'

We find it all too easy today, as at the time, to associate Baird's name with that of the company. Consequently, he attracts the accusations of sharp business practice: of overstating both the capabilities and the readiness of his television systems.

For product capability, the Baird Company fuelled expectations of what was going to be on offer and what television was capable of doing. This gave rise to wild imaginings and exaggerated claims on the part of the media. That it stemmed from the Baird Company more than anything contributed to Baird being branded a charlatan and a mountebank. This was exacerbated by claims of 'Television just around the corner'.

However, there is nothing unique or original in these practices. At the dawn of personal computing in the early 1980s, an advertisement for a small microcomputer, with its tiny usable memory (1 kilobyte) and fragile domestic quality construction, suggested through careful wording an ability to control a power station. Today we continually have the latest computer software announced by both major corporations and small companies long before the products are available to buy.

In his autobiography, Baird portrays himself largely as the innocent. For the most part, he claims to have been oblivious to the business implications, entrusting those issues to his team. This picture of innocence is difficult to believe, given his previous enterprises in business trading.

An Englishman, an Irishman and a Scotsman

Before his exploits with television, Baird tried his hand at various mass-product items. 'The Baird Undersock' was no more than unbleached socks direct from the manufacturer sprinkled with borax. 'Baird's Speedy Cleaner' was simply bulk soap. He knew how to re-package a product and sell it to the public.

Sharing in this enterprise was Captain Oliver Hutchinson (see Figure 7-2). Born in Belfast on 6th May 1891, Oliver George Hutchinson grew up in Ireland. The family home in Hillsborough, County Down in Northern Ireland, was built on profits from the family's successes in the emerging motor trade.[2] Hutchinson moved to Scotland where he worked for a time at Argyll's of Alexandria possibly at the same time as John Logie Baird. They definitely met after the war, in 1922, when they happened to be in competition for the soap business: 'Baird's Speedy Cleaner' versus 'Hutchinson's Rapid Washer'.

Fig 7-2. Oliver Hutchinson.
Courtesy of the Royal Television Society
RTS40-23

With the success of Baird's television demonstrations, Hutchinson joined Baird as partner, business manager and eventually joint Managing Director in 1925. Without Hutchinson's entrepreneurial approach, the name of Baird may not have been as well associated with early television as it is today. Hutchinson, later supported by the English financial journalist, Sydney A. Moseley, pushed hard to get recognition for television and to get a television service on air. The Englishman, the Irishman and the Scotsman succeeded, but not without a great deal of difficulty.

In mid-1928, the Baird Company applied to the Postmaster General for transmitting rights for a broadcast television station.[3] Baird already held an experimental licence, issued in August 1926, for two transmitters, 2TV and 2TW, that he and Hutchinson had bought. The first of these, 2TV, was installed on top of Motograph House and had an initial power of 250 Watts. This was raised to 500 Watts when the Company moved to Long Acre. Though 2TW was licensed for use in Green Acres, Harrow, no transmissions were ever made from there.[4]

If the Baird Company had had a licence to broadcast television programmes, it would have broken not only the monopoly by the BBC but also the policy set in 1922 by the Post Office. This policy made the BBC the sole public broadcasting service in Britain. The Baird Company wanted to get television on air to create a market for receivers, in a manner that was reminiscent of the action of broadcast radio manufacturers just a few years before. To get a clearer picture of what was at stake, it is worthwhile to look back to the beginning of the 1920s.

Radio

In the early 1920s, broadcasting by radio swept the United States of America in a spectacular fashion. In Britain, the Post Office as regulating authority received many requests for private companies to start a broadcasting service. The main reason was that those companies could sell receiving sets. Rather than having a chaotic situation with the possibility of virtually unlimited and uncontrolled broadcasting, the British Broadcasting Company was set up to be the sole provider of a radio broadcasting service. Agreement was reached with the Marconi Company, the holder of most of the relevant patents, and the BBC started broadcasting programmes on 14th November 1922, transmitting from station call-sign 2LO in Marconi House, London. In the following two days, two more transmitters started up: 2ZY in Manchester and 5IT in Birmingham. By March the following year, Newcastle, Cardiff and Glasgow commenced broadcasting, eventually followed by Aberdeen and Bournemouth in October 1923.[5]

As amply demonstrated in the United States, radio was turning out to be immensely popular. By the end of 1923, the number of licensed listeners in Britain had reached over half-a-million. Broadcast radio would have been 'the greatest thing since sliced bread', had it not been for the fact that sliced bread appeared much later, in 1930.

By November 1932, just after the start of the first BBC Television Service there were five million people in Britain licensed to receive radio.[6] One in five households had a valve radio set. This was what Baird had been after: a potential audience of millions who had already purchased radio receivers, with national radio coverage from the BBC on the medium waveband. In sheer numeric terms, in every country, radio was to completely dwarf television in all its forms for decades. Radio remained the 'Senior Service' until the 1950s.

Intentions made clear

In the Britain of 1928, the Post Office and the BBC still ruled the waves –
the radio wave spectrum, that is. Their assessment of Baird's television
went a long way to allowing other countries, mainly the United States, to
steal a march on Baird's hard-won 'achievements'. Based mostly on the
engineering aspects of Baird's then current offering, their assessment was
over-cautious and highly prejudging. There was far more at stake than the
amount, or the lack, of detail that could be seen. What truly mattered was
the provision of a broadcast service to experiment in not just the
technology, but in the programming. What the BBC seems to have ignored
was the tremendous public interest, fuelled by Baird's demonstrations ever
since 1925.

The prejudice against the Baird
Company seemed to have at its
focus the BBC's Chief Engineer,
Captain Peter Eckersley (see Figure
7-3). The BBC had little just-
ification in stopping the experimental transmissions: the 2LO
transmitter was in use for only an
average of 10 hours a day.[7] What
was even more alarming to the
Baird Company was that the BBC
was already supporting an experimental image transmission service,
based on a home product called the
Fultograph. This service broadcast
still facsimile pictures into the
home. In four minutes, the little
machine connected to a conventional radio receiver would deliver a
hard-copy picture of 14 cm by
9.5 cm (5½ by 3¾ inches). The

Fig 7-3. Captain Peter P. Eckersley.
From 'Broadcasting from Within',
C.A.Lewis, 1922

home Fultograph was a complete commercial failure, with less than 350
sets sold by July 1929 after eight months of out-of-hours broadcasting.
Fundamentally, it was far cheaper to wait until the next morning, go out,
and buy a newspaper.[8] In obstructing Baird and supporting the Fultograph,
the BBC had shown remarkable inconsistency.

When the journal 'Television' first appeared in 1928, it was used as a
vehicle, largely by Sydney Moseley, to lobby support for television
transmission experiments by Baird, which the BBC were said to be

blocking. With headlines emblazoning, 'Now then, Captain Eckersley', Moseley went for the BBC's throat by creating public sympathy and support for Baird.[9]

US Experimental Television Broadcasting takes off

Baird and his team had good cause for concern. At this time, the most prominent developer of television systems in the United States was Charles Francis Jenkins. Jenkins started broadcasting regular programmes from his broadcast station, W3KX, in Washington D.C. in July 1928. Through limitations in his equipment, he was transmitting merely shadowgraphs and silhouettes in vision only.[10] However, Jenkins had not been the first.

The General Electric Company had started broadcasting regular test transmissions over radio station WGY from Schenectady, New York in May 1928 using the mechanical system developed by Dr Ernst Alexanderson. These transmissions were initially only intended for the GE engineers rather than the public. Just four months later, on 11th September 1928, WGY hosted a 40-minute television play called 'The Queen's Messenger'. Even though it featured just two characters, it was the first dramatic programme on television.

WRNY opened up in New York in June 1928 with alternate vision and sound. As with Jenkins's system, the vision signal merely consisted of animated silhouettes. Television stations – all with experimental licences – were springing up in the United States. The aim of all these stations was simply to sell receiver-displays. In July 1928, the New York Times listed twelve stations.[11] By October, there were fifteen and in November twenty-one licences issued. Not all licences were used, but this period constituted the boom in mechanical television in the United States. What sprang up was not just a plethora of stations, but a miasma of different television formats. Before the situation got completely out of hand, the Federal Radio Commission called 'time-out' and put in place a regulatory framework. They allocated short-wave frequencies for more detailed television images and set standards of horizontally scanned 48-line and 60-line TV frames, at 15 frames per second.[12]

Getting on Air

In Britain, the intense lobbying appeared to be a means for the Baird Company to gain commercial, rather than any engineering, benefit. A BBC television service on the Baird system would create a market for Baird receivers. The engineering challenge lay in developing new, higher definition systems including cameras, transmitter and receiver systems. The Marconi and HMV organisations, when they first started in television, made progress in developing television without making a public issue of it and without the need to have a programme service whilst developing it.

We must not forget that the Baird Company was a commercial concern. Unlike the Post Office or the BBC, it needed to sell services and products to survive (see Figure 7-4). Unlike the established corporate companies such as Columbia, HMV and Marconi, it did not already have a pipeline of products to support long-term research. All the Baird Company 'eggs' were in the television 'basket'. It relied on financial backing of which there was a considerable but finite amount. Consequently, the Company was driven to get television out to the public as soon as possible.

Fig 7-4. The workshop at Baird's laboratories in Long Acre, London. (1931–32)
Courtesy of R. M. Herbert

The achievement of practical television in the 1920s pre-dated the development of the high bandwidth broadcast transmitter. As a result, television had to use the existing broadcast radio infrastructure. That meant transmitting on the medium waveband, and that entailed keeping the information content down to what would fit into one of the existing audio channels of just a few kilohertz. 30 lines in the television picture were about as much as the medium wave could stand. Though there was never any intention of having equal detail horizontally and vertically, such an approach would have needed a transmission bandwidth of 13 kHz. As such, the public would have to suffer slightly fuzzy pictures.

After over a year of demonstrations and acrimonious discussions, the Baird Company received an assurance from the Postmaster General in late March 1929. He said that the Baird Company should use the BBC's existing medium-wave transmitter outside normal transmitting hours for experimental television broadcasts.

Throughout this period, it is difficult to see any special development by the Baird Company that caused the Postmaster General to grant permission. Simply, the BBC and the Post Office relented from what amounted to a bitter and prolonged attack. The major opponent to Baird in the BBC, Captain Eckersley, resigned from the BBC in 1929.

The Baird Television Service

On 30th September 1929, the BBC started an experimental broadcast service using Baird's 30-line equipment and studio facilities operated and financed by the Baird Company. The BBC provided only the facilities for transmission, at a fee. At the time, the BBC could only offer one transmitting channel, through the London transmitter call-sign 2LO (see Figure 7-5). A single transmitting frequency meant that either the vision, or the sound, but not both together, could be transmitted at any one time. It seems amazing today that such an idea could have been accepted. We can understand why it was when we think of this more as a radio programme with illustrations.

The inaugural programme, with speeches, songs and a comedy turn, was more of a radio programme than television. As each of the guests was introduced and before they spoke or performed, the audio was switched over to the video signal. Anyone with a Televisor display would in theory be able to see on his or her screen the person who was about to speak or perform. The vision signal was transmitted for only two minutes at a time, before going back to audio. Considering the picture had to be manually synchronised for correct framing, and considering not many 'lookers-in' would have had a tremendous amount of experience in doing that, two minutes would not have left any time to sit back and enjoy the transmission.

With only one operational transmitter, there was no alternative to this state of affairs. There was also nothing particularly unusual in this, in that almost all the experimental television broadcast stations in the United States operated in this way, time-multiplexing vision and sound through the one transmitter. The reasons for this situation persisting for any length of time were the novelty of having vision by 'wireless', the public's interest in television and the difficulties of using two separate frequencies for vision and sound transmission and reception.

Despite 2LO's low power (only 3 kW at that time), the television images were received over remarkable distances and attracted enthusiastic reception. The split sound and vision was a problem that could only be resolved by an additional transmitting frequency. On 31st March 1930, six months after the inaugural transmission, the BBC offered simultaneous vision and sound programmes through its twin wavelength transmitter at Brookmans Park, on the north side of London, which had opened for service in October 1929.

Fig 7-5. The BBC's 2LO transmitter, manufactured by the Marconi Company.
Courtesy of the BBC, RTS38-103

The world's first television service with regular scheduled programmes was not the BBC, but Baird Television, the name that the Baird Company was to adopt in May 1930. The Baird Television Development Company, to give it its full title, funded the complete service, from camera and studio to performers and producers (see Figure 7-6). They leased the Post Office line to the BBC transmitter and even paid the BBC £5 per half-hour for the use of the transmitter.[13] Having the signal transmitted on the BBC frequencies leant tremendous credibility to the Baird Company.

In what must be thought of as the Baird Television Service, the most notable broadcast was that of the play, 'The Man with the Flower in his Mouth', by Luigi Pirandello, broadcast on 14th July 1930. Opening to the strains of Carlos Gardel singing 'El Carretero' (the ox cart driver), and introduced by the co-producer Lance Sieveking, the simple play became a defining moment in British Television, though its reviews at the time were somewhat mixed.[14]

In 1967, the Inner London Education Authority enacted the most authentic version of the play since it had been broadcast in 1930. Sieveking returned with original props, captions and incidental music. He provided the essential voice-over, just as he had done for the first performance. The authenticity was secured by Bill Elliott, who had modified one of his two precious Baird Televisor receivers to operate as a camera. He recorded extracts from the play onto reel-to-reel tape, sound on one channel, 30-line vision on the other. Despite the distortion caused by being directly recorded onto tape, the video signal was remarkably clear.

Fig 7-6. Outside Baird's laboratories and studios at 133 Long Acre, London, an outside broadcast is underway on 8th May 1931. On the left is a caravan within which is a mirror-drum camera. Cables lead off to the upper floor of the building. The caravan is angled to the pavement and would capture the full height of the gentlemen on the right. The person on the extreme right is Sydney Moseley.

Courtesy of the Royal Television Society RTS 36-41

The Baird Company was to all intents and purposes running a National Television Service, albeit experimental, out of its own pocket. The costs of operating this service together with costs from years of development and demonstrations were simply not being offset by any appreciable income. By 1931, an extremely tough year for any company in Britain, the Baird Company was crippled financially. Sydney Moseley helped in arranging a take-over of the Baird Company by Gaumont British Pictures Corporation, which occurred in January 1932. The name of Baird was retained for the company name; Baird was undeniably a legend and a leading light in television. With the Baird Company though, the new masters sought to create a more financially aware organisation.

The First BBC Television Service

In the third year of television broadcasts from Baird's studio in Long Acre, the Baird Television Development Company reached agreement with the BBC on the future of 30-line television. From August 1932, the BBC would take on total responsibility for television broadcasting. The Baird Company loaned out its camera system and studio equipment (at no charge) and transferred two engineers to the BBC. The equipment and the Baird engineers – Desmond Campbell and Tony Bridgewater – were installed in a basement studio, Studio BB, in Broadcasting House. These premises were brand-new, officially opened on 20th May 1932. Throughout the hot summer of 1932, the small sound studio was transformed with the installation of brand-new 30-line equipment to meet the challenges of the new studio. Designed by J. C. Wilson and built by Messrs B. J. Lynes Ltd, the camera system (see Figure 7-7) was an excellent piece of mechanical engineering.[15]

Unlike the fixed camera system used up to that time in the Baird studios, this camera could be 'panned' from side-to-side. Adjusting for tilt amounted to positioning metal jaws in front of the camera. These jaws defined the video frame. Adjusting the jaws vertically altered the video timing, with lines occurring early or late. This could upset the delicate synchronisation at the receivers. Like all Baird's camera systems since 1928, this was not exactly a camera, but a 'flying-spot' projector. A Zeiss 10 Amp carbon arc light created an intensely bright beam of light. This shone onto a 20 cm wide drum with 30 mirrors bonded around its circumference.[16] Each mirror reflected the intense beam out into the studio. At each turn of the mirror-drum, the 30 mirrors created a rectangular array of 30 lines sweeping over the object in the blacked-out studio.

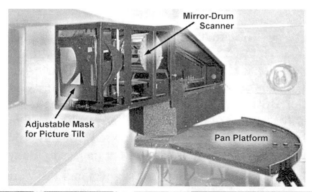

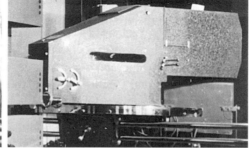

Fig 7-7. The BBC's 30-line television 'camera' projector in use from 1932 to 1935. At top is the projector in Studio BB in Broadcasting House. Below left and right is the projector at BBC Langham Place in March 1934 and April 1935 respectively.
(Top) Original from Reyner, 'Television Theory and Practice', 1934
(Below) Courtesy of the BBC, (left) RTS37-85, (right) RTS37-86

There was no viewfinder to this camera; the operator could see the scanned area projected into the studio and could adjust the focus by looking at how well defined the line structure was. The camera-projector was positioned where a conventional camera would have been. Where we would expect the lights to be, there were photoelectric cells. The studio used four movable clusters of cells that could be positioned just as if they were lights, and with the same overall effect. Lighting control amounted to varying the output from the photocells. Turning down the output from one cluster of cells was just like turning down the brightness of a studio light. Although the reversal of roles of lights and pick-up devices did not matter to the final picture, there was one serious drawback. The performers were operating in a pitch-black studio. With no possibility of reading 'prompt-boards', they had to memorise all their lines. They could see the vertical area illuminated by the camera-projector, and the light dimly reflected off that area into the rest of the studio. When they were in shot however, all they could see was the intense arc light blasting straight into their eyes, at twelve and a half times a second.

First Light

The first BBC TV programme was broadcast on 22nd August 1932 at 11:02pm and lasted just 35 minutes. With a few words from BBC Director of Programmes, Roger Eckersley, about the new service, John Logie Baird thanked the BBC and expressed hope for the future of television as a public service. The BBC producer, Eustace Robb, had arranged light entertainment for the rest of the show, and those who tuned in watched and heard dances and songs from Betty Bolton, Fred Douglas, Louie Freear and Betty Astell. The programme closed with four dances by Betty Bolton.

Critics the next day praised the new service, and the work of Robb, as being '... the latest modern miracle of science with a future assured...'.[17] The service continued with four half-hour shows per week, on each weekday except Wednesday at 11pm with Robb as producer. As with the 30-line television service itself, the work of Eustace Robb was overshadowed by the achievements of the 405-line television service and producers such as Cecil Madden and Dallas Bower. In 1934, Robb reviewed his first eighteen months of work as a television producer.

> 'In this short time, the technique of programme presentation has progressed rapidly, and artists have more latitude for movement. They can appear in what is called the "long-shot", which shows the entire figure, though with unrecognisable features, and be focused up to the close-up when their features become clearly recognisable. Dancing has naturally played a big part in television ... But dancing is not the only new form of entertainment to be brought by television to Broadcasting House. There have been boxing matches, mannequin parades, acrobatic turns, animals from the zoo, performing animals, musical comedies and revues.'[18]

In January of that same year, television's big brother, radio, sought to reclaim Studio BB for its original purpose – as a rehearsal studio. Broadcast radio was still expanding as receiving sets became cheaper and the demand increased. The 30-line service was really only guaranteed to March 1934 and thereafter, could be terminated with 6 months notice. The BBC gave that notice in September 1933 in preparation for closure on schedule. However, it subsequently decided to move the equipment to a distinctly larger studio at BBC premises in Portland Place, just a short walk from Broadcasting House (see Figure 7-8). The BBC could so easily have axed the service, yet it extended the facilities and the service, and allowed the camera system to be upgraded. The face-lift was in the form of more sensitive photocells from the Oxford Instrument Company. Progressively introduced from 1934 onwards, these cells allowed better signal quality in the larger Portland Place studio. After the last programme from

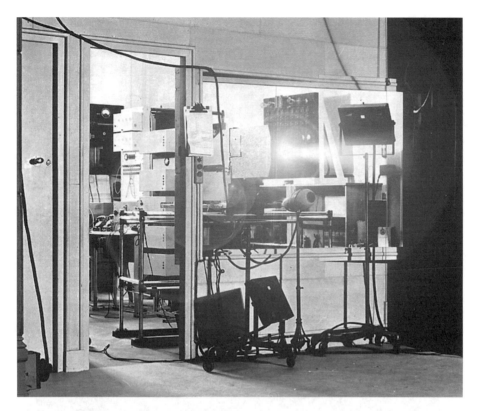

Fig 7-8. View from the studio floor at BBC Langham Place looking into the Control Room. The light from the 'camera' scanner is here simulated, modelled on a similar but poorer quality photograph. In use, the studio was in the dark, lit only by the flickering light from the scanner. Off to the right, live music would be performed behind a light-proof curtain.

Original courtesy of the BBC, RTS 37-97

Broadcasting House on the 16th February, there was a busy but uneventful move of all the equipment into the first floor suite of a fine Regency house in Portland Place. The service re-started from the new studio on the 26th February 1934.

Progress on 30-line Quality

We tend to think today of 30-line television as being particularly poor quality. We do not fully appreciate that there was a continual progress in quality, right from Baird's experimental days to its demise in 1935.

Baird's early Nipkow disc cameras used conventional flood lighting of the subject. The cameras had to be huge and were difficult to make. Consequently, the errors, primarily in positioning the lenses on the disc, gave the images a decidedly ragged look. The sensitivity of photocells and

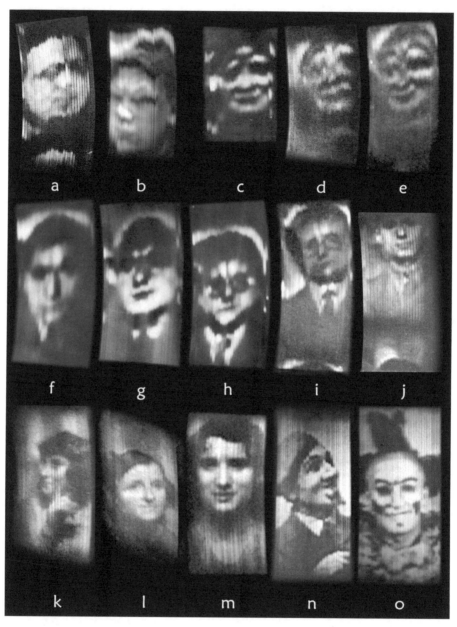

Fig 7-9. The progression of 30-line television quality. Though not all pictures were taken in similar circumstances, this indicates the development in quality over nearly ten years. **a)** first off-screen photo, 1926, **b)** early 1926 image (cf Fig 3-3), **c), d), e)** 'Stookie Bill' dummy heads in 1928, **f)** G. B. Banks with pipe or cigarette (1928), **g)** unknown (1928), **h), l)** Philip Hobson 1928, **j)** unknown subject with picture offset (1928), **k), l)** unknown females, **m)** Mabel Birch (1931), **n)** 'Harry Lauder', **o)** John Rorke 3rd March 1933
Courtesy of R. M. Herbert, P. Hobson, the Royal Television Society and the NMPFT

Fig 7-10. A 30-line television mirror-drum. Each mirror is canted slightly to place each line of the picture next to the previous. The step from lines 1 to 30 occurs just below the reflected glare. The precision mechanical components for Baird's systems were for the most part sub-contracted to B. J. Lynes Ltd.

Courtesy of R. M. Herbert

the performance of amplifiers also contributed to poor quality (see Figure 7-9). When Jacomb brought the flying-spot technique to Baird's team in 1928, he created a major overall improvement to the quality of the picture.

Right from the time that television broadcasting began, the camera system improved significantly. The fixed field-of-view and *staring-eye* of Baird's Long Acre system was a major problem for staging entertainment. The BBC mirror-drum system of 1932 onwards meant that the staring-eye could now be panned and adjusted for tilt (see Figure 7-10). The essentials of the performance could now be captured very effectively. If a performer moved, the camera could follow the action. The ability to pan the camera was probably the single biggest step forward for 30-line programme production. Though the BBC cameras and transmitting system was improved throughout its life, the greatest improvements came in the design of new display devices.

In January 1934, Solomon Sagall, Managing Director of television receiver manufacturer Scophony Ltd, described the improvements.

'Compare "30-line" pictures as seen a few years ago, even those of twelve months ago, with what is offered now. What a vast improvement! One would never have expected that such results could be obtained from what is, after all, only a 30-line picture.'

'This tremendous improvement is, of course, also due to the greater efficiency of modern television receivers. The difference between a 30-line picture of a few years ago, the size of a postage stamp, as seen on a Nipkow disc, and a BBC picture received on a Scophony 30-line rotating echelon or on a Baird mirror-drum, is like that between a doll with eyes made to move and a strong healthy infant.'[19]

Though there is undoubtedly an advertisement embedded in Sagall's article, his observation regarding the improvement going from Nipkow disc to mirror-drum is just as valid for receivers as it was for the camera systems. The sale of new types of receivers was not going well from 1933 onwards. The Bush mirror-drum receiver (see Figure 7-11), manufactured and distributed exclusively for Baird Television Ltd, was an elegant piece of walnut furniture standing 4ft 1in (125 cm).[20] It housed a mirror-drum projector that created a bright image on a 9-inch by 4-inch (23 cm by 10 cm) screen. This was to be the next generation 30-line 'Televisor' complete with a single radio receiver for the vision channel. It was assumed that buyers of this expensive 75 guinea (£78.75) vision-only receiver would already have a conventional radio to pick up the sound channel. Like the Scophony receivers, the set never sold. Despite the undoubted improvement in quality over the Nipkow disc, the impending demise of the 30-line service and having only two programmes to watch per week secured the fate of all new 30-line equipment.

On top of the technical development of the television camera system, the BBC's broadcast distribution system and home display systems, equally impressive steps were being made in programme production. In an editorial comment made in late 1934, 'noticeable improvements have been made in studio technique and presentation even during so short a period as the last three months.'[21]

The New Wave

Whilst the 30-line service was still operating, every television development company around the world, even the Baird Company, was developing higher definition systems. Initially the focus was on better mechanically scanned systems. In 1932 in Britain, EMI, Marconi and Baird were all developing camera and display systems capable of more than 100 lines. The increased information content needed transmitters capable of carrying that information, and that led to progressively higher radio frequencies and into Ultra-Short-Wave (above 30 MHz).

In the same month that the BBC broadcast what they hailed as the World's First Television Revue, 'Looking In' in April 1933, the Baird Company were demonstrating a 120-line system using an electronic display. Two months later, the Baird Company moved out of Long Acre and took on larger premises in the South Tower of Crystal Palace. To run all technical development, the Baird Company, under the new management of Gaumont British, had recruited Captain A. G. D. West. West had come from the Gramophone Company's (pre-EMI) research and designs department and before that, between 1923 and 1929, he had been chief research engineer with the BBC.

West's background may explain his own hiring, Jacomb's departure in 1933 and Baird's removal from the mainstream technical development activities of the company. West had known of Baird and his work back in the Hastings days. He was a great supporter of Baird's work and his continued contribution to the Baird Company.[22] Nevertheless, the company was focused on revenue generation and, from 1933, was in heavy pursuit of providing all the equipment – from cameras to domestic receivers – for the new high definition television service.

In 1931, West had made a visit to the United States and became aware of the developments in electronic television in RCA. In fact, his report of that trip was instrumental in EMI scaling up their

Fig 7-11. The Baird mirror-drum receiver.
Courtesy of the Author
From the collection of the NMPFT

television development efforts. By 1933, EMI had become a major threat to the Baird Company. In the space of a few years, EMI had gone from nothing to outstripping the Baird Company in the quality of its demonstrations of television.

End of the Line

The initial BBC 1932 weekly schedule of one programme per night, on each of four weeknights did not last for long. In April 1934, the service was cut back to two programmes per week, though the programme duration was extended to 45 minutes each. The times and days were changed in October 1934 to include Saturday afternoons, and then again in April 1935, to Mondays and Wednesdays after 11pm.[23] The cutback in programmes may have been partly due to the great developments and demonstrations of higher definition television. These developments drove the Government to form a television committee, under Lord Selsdon, to determine the future path for television in Britain. In January 1935, they issued their recommendations – that as soon as practicable, the BBC should start a high definition broadcast television service.

Strangely, the BBC 30-line service continued on, eventually finishing on 11th September 1935 with the 11pm night-time show. The announcer, Geoffrey Wincott, read out, 'some of us who have followed the television programmes for the past three years will feel somewhat sad that we are looking at the last programme on 30-line television'.[24] Several of the then well-known performers who had made television happen turned up for this last-ever programme. The first performer was Cyril Smith, the main classical accompanist since the start of BBC Television over three years before. He played Chopin's Polonaise in A Flat for solo piano. Several other acts followed in a full programme of entertainment.

> 'It fell to Lydia Sokolova to end the programme. The great ballerina had flung herself to the floor in a climax to the Baccanale (Moussorgsky) when the light of the scanner was dimmed for the last time ... "And that, ladies and gentlemen, is the end of the last television programme on 30-lines, so may we say Goodbye"'[25]

Intermediate Film – the Intermediate Answer

In an attempt to maintain a competitive position against EMI and other emerging competitors, the Baird Company adopted a German-developed idea for high definition television using film.. The method first appeared in a patent by Hartley and Ives in the USA in 1927.[26] The approach provided a high definition image without recourse to new developments. The idea was somewhat bizarre, but reflected the desperate need for a high-resolution television camera. The scene was filmed on cine-camera and the exposed film was developed and processed immediately within the mechanism. The film was scanned by a mechanical flying spot system, giving the television picture. Initially the picture had 180-lines (see Figure 7-12), but was extended for the 1936 broadcast service to 240-lines. Unlike the German approach, the film was not re-cycled, limiting programmes to the length of a film canister, around 20 minutes. With the 'pipeline' delay through the processing tanks, the audio had to be recorded onto the same film. The system was no more than a stopgap until the Baird Company could get an alternative. This was called the Intermediate Film process, as there was a filming stage between the studio performance and the television image output.[27]

The Baird Company at Crystal Palace worked hard into the mid-1930s to bring together the components of a high definition television service. Like the emerging competition, they developed the vision systems, the VHF transmitters and receivers, and the electronic displays necessary for a full television service. They had multiple solutions for the camera, with the type depending on what it was to be used for: the Intermediate Film process for studio work, a flying-spot high definition camera for a tiny single person studio for continuity announcing and, shipped over from the United States, the Farnsworth Image Dissector camera.

Fig 7-12. Madeleine Carroll. On the left is a single frame from a test film loop used by Baird engineers for development of the Intermediate Film system. The off-screen 180-line result on the right shows a 'work-in-progress' view complete with horizontal dark streaks. These are reportedly caused by dirt in the apertures.
Courtesy of the Royal Television Society RTS38-67 (left) and R. M. Herbert (right)

Fig 7-13. Jane Carr with high contrast make-up for 30-line television (centre) and off-screen results in the studio (left and right), 15th November 1932. The face is whitened and blue-black is applied to eyebrows, eyelashes, side of nose and lips. Heavy white is applied between eyebrows and eyelids.

Courtesy of D. R. Campbell

All-Electronic Television takes Shape

At the newly formed EMI Company at Hayes in Middlesex, work was underway on providing the elements of a full television service. At first, EMI seriously considered the mechanical solution, basing all their early experiments on a mechanically scanned camera but with a cathode-ray tube for display. Their results, particularly with help from the Marconi Wireless Telegraph Company in transmitter technology, were deeply impressive and, in demonstrations to the BBC and the Post Office, were considered vastly superior to what the Baird Company could produce. The demonstrations led to a proposal from EMI to the BBC to provide all the equipment necessary for a television service. EMI intended the BBC to operate the service. Like their competitors, they were interested primarily in the inevitable market for television receivers.

EMI were taking television extremely seriously. They had ready access to the patents owned by the Radio Corporation of America (RCA), which described the all-electronic camera tube or *iconoscope*, invented by Vladimir Kosma Zworykin. By 1932, William F. Tedham and J. D. McGee of EMI had successfully built and run an independently developed electron tube camera based on the principles set out in the US patents for the iconoscope. Though experimental, Tedham and McGee's efforts showed that a practical solution was imminent.

Whilst these rapid developments in higher definition television were underway, the BBC continued with its 30-line service, developing and gaining experience in techniques for making live television programmes (see Figure 7-13).

Television in Transition

Developments and demonstrations in electronically scanned television led to the BBC's 30-line Television Service being stopped in 1935. By this time, low definition television had virtually national (almost European) coverage and had an audience of somewhere between 8,000 and 15,000. The viewing public were keen for transmissions to continue, even in parallel with a high-resolution service. After the service closed and the last programme was broadcast, these 'lookers-in' found that they had nothing to watch – their 30-line television receivers had become obsolete.

Unlike the much-vaunted advent of digital television in 1998, the transition from 30-line to the new high definition service was not an enhancement but a total revolution. We must remember that 30-line television was designed to use the existing radio channels intended for audio broadcasting. The BBC's 30-line studio equipment was mature technology. It had used its existing audio distribution channels and radio frequencies for vision transmission, leaving the public to buy or even build their own receivers.

Fig 7-14. Baird Television receiver Model T13 (1937).
Courtesy of the Royal Television Society RTS37-68

In sharp contrast, a totally new infrastructure supported the high definition system. Virtually everything had to be developed from scratch – cameras, cables, distribution equipment, routers and transmitters. For the consumer, they required an antenna, a receiver and display, all of which were brand new (see Figure 7-14).

The investment was enormous but the time was right and the public (and hardware manufacturers) were crying out for a full television service. The potential returns from such a service made the investment look secure, despite the prospects of another War.

Trial by Television

When BBC television test transmissions started in 1936 from the Radiolympia exhibition in London, the price of receivers, full of the latest technology left the public far behind. Initially, television sets had to be dual-standard: the choice between the Baird Company's totally new 240-line progressive scan system and rival Marconi-EMI's 405-line interlaced system was to be resolved on-air. Much like the start of the BBC's digital service in 1998, hardly any of the public had the new receivers to watch it.

In its first few months in late 1936, the new service was far more experimental than its 30-line predecessor. There were however increasing concerns over the Baird system. The Baird Company board played down the poor performance of their system, recognising that the money was going to be made on the television receivers rather than on just one set of broadcast equipment. In January 1937, the 240-line BBC service using

Fig 7-15. Rescued after the Crystal Palace fire from the remains of the Baird Company's laboratories was a Farnsworth Image Dissector, melted by the heat.
Courtesy of the Author, from collection of R. M. Herbert

Baird Company equipment was dropped in favour of Marconi EMI's 405-line television. Marconi EMI's all-electronic solution was simply better overall. The last programme on the Baird system was broadcast on 30th January 1937. This was the second blow in two months. On 30th November 1936, the Crystal Palace had been destroyed by fire, and with it the Baird Company's laboratories, studios and everything stored within (see Figure 7-15).

Viewers outside the London area, who switched off their 30-line receivers for the last time in 1935, had to wait more than 15 years for television to return. It took until 1952 for coverage to reach Scotland and Wales and 1953–54 for prices of receivers to become affordable to the average working family. [28]

Though the Baird system had been inferior to the Marconi EMI system for broadcast television, there was one area in which the Baird system excelled – airborne reconnaissance.

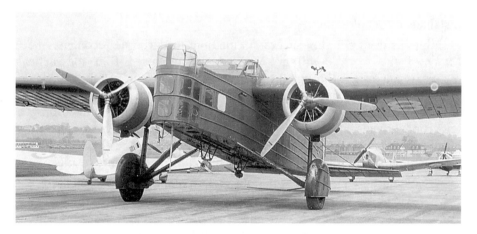

Fig 7-16. The Marcel Bloch 200 bomber of the French Air Force housing the Baird airborne reconnaissance equipment. The antenna can be seen hanging down from the nose of the aircraft, in front of the power generator (with small two-bladed propeller).
Courtesy of R. M. Herbert

Airborne Reconnaissance – 1937

In 1937, the French Air Ministry awarded contracts to the Baird Company and to Marconi EMI. These were for airborne reconnaissance systems to an identical requirement, to be fitted for trials onto Marcel Bloch 200 twin-engine bombers (see Figure 7-16). Both companies implemented systems that were adapted from their offerings for broadcast television, and developed further to meet the contractual requirement of 400-line progressive-scan resolution from an aircraft. The use of progressive scanning was important. Although interlaced-scanning was useful for reducing apparent flicker in the studio, it was no use in the continuously scrolling picture from downwards-looking airborne reconnaissance.

Marconi EMI found they had problems; their iconoscope camera tubes were picking up vibration from the engines causing their images to be heavily degraded. The Marconi EMI solution relied on what the camera tubes could resolve.

In contrast, the Baird system, based on the Intermediate Film process, used a cine camera for capturing the scene. The film was processed on board and passed through a 'tele-cine' apparatus. This created a television image, which was transmitted to the ground station a maximum of 40 miles away. In terms of success, compared with the trial for BBC Television, the Baird and Marconi EMI solutions swapped roles. There were distinct advantages in capturing the imagery using high resolution film as the storage medium, and creating a real-time television view. This view could be monitored on board the aircraft and transmitted to the ground station for instantaneous exploitation.

Film has been a highly successful medium for airborne reconnaissance. Many of the world's military air forces have used film for tactical reconnaissance right throughout their history. The electron tube television camera, though best for broadcast television, did not give as high resolution as film. Before the 1960s, there was no possible means of recording a television signal on-board the aircraft. When airborne video recorders eventually appeared, they were at the mercy of the harsh military environment and required elaborate protection for best quality results.

What comes as a surprise is that the Baird reconnaissance project for the French Air Force was functionally *identical* to film-based reconnaissance systems in operational use with many air forces around the world some fifty years later. Just like the Baird system in 1939, the imagery was recorded onto film on board the aircraft for later analysis, was monitored on board for quality checks and was sent directly via a data-link to a ground station for immediate exploitation. The primary difference is that, in 1939, the French bomber gathered its imagery at an altitude of 5,000 feet whilst lumbering along at the stately speed of 150 mph (240 kph). It would have been a sitting duck. Ray Herbert was a Baird engineer working on the reconnaissance project. 'During the trials in July 1939, I remember saying to the RAF sergeant outside the hangar one morning, "Do you think there is going to be a war?" He said, "I hope not for your sake, you wouldn't last five minutes in that thing."'[29] War broke out a few months later, and the project went no further.

Fig 7-17. 120-line Colour Television in 1938. Paul Reveley (left) is pointing the colour television camera system at Miss Sabrey (seated) daughter of an Egyptian diplomat, whilst John Baird stands by.

Original courtesy of R. M. Herbert

Baird's Colour Television

From 1933 onwards, John Logie Baird had far less to do with the primary activities of the Baird Company. The future of the Baird Company was not to be in Baird's hands. Baird set up his own laboratory, with around five assistants, all funded by the Company. He set about exploring other aspects of mechanical television almost independently of the Company, but notably in the area of cinema television in colour. By late 1936, he had demonstrated 120-line monochrome television transmitted from Crystal Palace and projected onto a large screen in the Dominion Theatre in the West End of London.[30] A year later, in February 1938, he gave a demonstration of a two-colour system on 120 lines, again projected onto a large screen at the Dominion.[31] These large screen demonstrations took place at a time when 405-line electronic television was well established; seeming to support the argument that Baird stuck with mechanically scanned television for rather too long.

On the contrary, throughout the war years, Baird developed and demonstrated a two-colour television system using an electron tube colour TV display. The colour was frame-sequential: two frames (orange-red followed by blue-green) were sent one after the other so rapidly that the eye saw a full-colour image. Frame sequential colour television was used commercially in the USA prior to the NTSC system of the 1950s and more recently used in broadcast television from the Apollo missions to the moon. There a colour filter wheel spinning in front of a monochrome television camera generated the colour image.

Back on 16th August 1944, Baird gave the world's first demonstration of a fully electronic colour television display to the press. His 600-line colour picture used triple interlacing, taking six scans to build each picture.[32] This was an astonishing personal achievement for Baird considering he was supported only by a glass-blower, and an engineer, E. G. O. 'Andy' Anderson. The *Telechrome* colour display tube was never developed further. Despite the success of colour television in the United States in the 1950s, the British public had to wait until the late 1960s for a progressive rollout of a colour TV service across the nation – a full 25 years after Baird's Telechrome system There were to be no more developments from the man who had dedicated his life's work to television. After a period of increased illness, and a stroke in February 1946, he died in his sleep on 14th June 1946.

[1] BAIRD, J. L.: 'Sermons Soap and Television' (Royal Television Society), 1988, p91

[2] MCLARNON, O.: Private communication with the author, Jan 2000

[3] BURNS, R. W.: 'British Television, the Formative Years' (Peter Peregrinus Ltd), 1986, p103

[4] HERBERT, R. M.: Private communication with author, 1st Feb 2000

[5] LEWIS, C. A.: 'Broadcasting from Within' (Newnes), p28

[6] BRIGGS, S.: 'Those Radio Times' (Weidenfeld & Nicolson), 1981, p54

[7] EXWOOD, M, 'The Births of Television', *The Radio and Electronic Engineer*, **46**, No 12, p632

[8] BURNS, R. W.: 'Wireless Pictures and the Fultograph', *IEE Proceedings*, **128**, Pt.A, No.1, Jan 1981

[9] *Television*, Feb 1929, p11

[10] UDELSON, J. H.: 'The Great Television Race' (Univ. of Alabama), 1982, pp50–51

[11] 'Stations Licensed for Television', *New York Times*, 21st July 1928, p16, reported in UDELSON, *ibid.*

[12] UDELSON, J. H.: *ibid,* p45

[13] EXWOOD, M.: *ibid,* p21

[14] 'El Carretero' written by Arturo de Nava, performed by Carlos Gardel, Recorded in Paris 11th Oct 1928. Information supplied by David BELLER, Israel

[15] BRIDGEWATER, T. H.: 'Just a Few Lines' (British Vintage Wireless Society), 1992, p8

[16] BRIDGEWATER, T. H.: 'Just a Few Lines' (British Vintage Wireless Society), 1992

[17] *Daily Telegraph*, 23rd Aug 1932

[18] *World Radio,* 2nd Mar 1934, reported in BRIDGEWATER, *ibid,* p12.

[19] SAGALL, S.: 'Television in 1934', *Television*, Jan 1934, p4

[20] ANON.: 'Manufacture of new "Televisors" ', *Television*, July 1933, pp241–242

[21] ANON.: 'Comment of the Month', *Television*, Nov 1934, p475

[22] BAIRD, J. L.: 'Sermons, Soap and Television' (Royal Television Society), 1988, p129

[23] ANON.: 'Comment of the Month', *Television*, Nov 1934, p475

[24] BRIDGEWATER, T. H.: *ibid*, p18

[25] BRIDGEWATER, T. H.: *ibid,* p18

[26] HARTLEY, R. V. L. AND IVES, H. E.: 'Electro-optical transmission system', U.S. Patent 2,166,247, 14th Sep 1927, British Patent 297,078 19th Mar 1928

[27] HERBERT, R. M.: 'The Baird Intermediate Film Process' (Journal of the Royal Television Society), **24**, No 3, May–June 1987

[28] BRIGGS, A.: 'The BBC: The First Fifty Years' (Oxford University Press), 1985

[29] HERBERT, R. M.: Interview with Author, Feb 2000.

[30] ANON.: 'Baird Big-Screen Development', *Television & Short Wave World*, Jan 1937, pp26–28

[31] ANON.: 'Baird Colour Television', *Television & Short Wave World*, Mar 1938, pp151–152

[32] HERBERT, R. M.: 'Seeing by Wireless' (R. M. Herbert), 1st edn, 1996, p26

8   It's All in the Groove

'The quality of television programming today tells
us there is no way television had a father...'

H. Selwyn-Mobberley, Glasgow 1974

The First 'International' Television Service

The demonstration of 30-line television reception across the Atlantic in February 1928 indicated that the low definition signal would travel well. With Baird's television service from 1929 and the BBC's service from 1932, various experimenters found that the television programmes from the BBC transmitters could be received over vast distances. There was nothing magical about this; it had to do with the natural long-distance propagation of medium-wave transmissions, especially in the hours of darkness.

Coverage often extended well beyond the vicinity of the transmitters, spanning almost all of Britain. The letters pages of 'Television' magazine throughout the 30-line era contain many instances of reception from Iceland to North Africa and Madeira. By the accounts of the 'lookers-in', they were able to make out a considerable amount of detail, much more than we can see today on these disc recordings. The reports are subjective and occasionally over-enthusiastic, but the ability to receive recognisable television pictures across Europe from the BBC is well documented.

Some amateurs even went so far as to build dual-standard Televisors to display the television pictures coming from Germany as well as Britain. The German standard called for lines scanned horizontally. If this had been electronic television, building a display system for both horizontal and vertical scanning would have been a complex affair. The simplicity of the Nipkow disc meant that there only needed to be two sets of holes defining where the picture was going to be and two viewing ports; one to look at the British format to one side of the disc and the other to look at the German format at the top of the disc.

Audience Numbers

In comparison with the millions of radio listeners across Britain, the television audience was miniscule. Unfortunately, there are no reliable audience figures to say how small. Five thousand official Baird Televisor receiving sets were sold and there were thousands of kits sold. However, the simplicity of the Televisor meant that a capable amateur could make a display from components. The engineering hobbyist magazines were full of 'How to' constructional projects for home Televisors. By 1933–34, top-end estimates ranged from '10,000 to 13,000' to '15,000 to 20,000' viewers or 'lookers-in' watching the BBC programmes throughout Britain.[1,2]

In the 1930s, the BBC did not consider any special need for a separate television licence, relying on the radio receiving licence. Consequently, there were none of the later Television Detector Vans to trap unlicensed viewers and contribute to understanding the size of the television audience. Estimates that are more conservative suggest that the television viewing audience in Britain was around 8,000.

Those amateurs felt as if they were pioneering a new frontier. They were in at the start of television, radio was coming out of its infancy and electronic components were becoming affordable to the experimenter. The introduction of sound for movies had occurred in 1927 with Al Jolson as 'The Jazz Singer'. The technology for the movies had been developing and maturing throughout the early 1930s.

Almost everywhere people turned, they saw technological progress. In aviation technology, speed records were continually broken and airlines offered journeys that had been just a dream a decade before, for those rich enough to pay for them. Even in the home, affordable electrical goods such as washing machines and vacuum cleaners helped relieve some of the drudgery from the housewife. Kitchen drudgery was alleviated by the availability of pre-packaged food, with sliced bread entering folklore as a reference against which 'important things' were compared.

The Amateurs

From comments in the pages of the 'Television' magazine, some radio amateurs were actively involved in trying to record the 30-line vision signal on their domestic recorders. In February 1932, R. C. Kaye of Huddersfield wrote describing his home-built system and his early attempts to set up for recording vision. 'I have recently been experimenting with home recording, with a view to making records of television transmissions.' He said that he had not tried it out.[3]

In a letter published in August 1932, F. G. R. Palmer of Sunderland claimed to have viewed the results of his recording of television received from Berlin.[4] Berlin was yet another target for the television buff at that time. At least one other enthusiast, in Aberdeen, regularly received and viewed test transmissions from the heart of Germany.

By 1934, H. J. Barton Chapple had written an article entitled 'Canned Television'.[5]

> 'Many have voiced the suggestion that it should be possible to record the television signals broadcast by the BBC on some permanent or semi-permanent device which could be used to furnish images in the home at any convenient time.'

Fig 8-1. A fanciful and impractical concept of 'Bottled Television', mixing gramophone recordings with high definition television.
From 'Practical Television' magazine, July 1935

Barton Chapple described how even studio recording techniques could not capture and reproduce faithfully all the frequencies of a 30-line video signal (see Figure 8-1). This, he explained, was the reason that no such discs were available to buy. However, he strongly encouraged experimentation.

> '...but this should not in any way prevent anyone from carrying out their own experiments, provided they appreciate that the results to be expected will not live up to "one hundred per cent".

I have done this work several times myself, using one or two of the home-recording devices which have been on the market from time to time.'

Barton Chapple made these comments in a draft form of his article. By the time his article had appeared in print, there was no mention of him having tried out recording.[6]

Domestic Audio Recorders

The home recorders referred to by Barton Chapple were all in the category of domestic equipment using gramophone recording. Tape recording became practical very much later, starting in the 1950s with the first reel-to-reel machines. In the early 1930s, the technology for recording in the home covered the whole gamut. The capable enthusiast could go out and purchase simple adapters for existing gramophone decks, all the way up to complete combination units that could record audio in synchronisation with a cine camera and projector. Home 'talkies' were quite practical, if you had the money to spend and you did not mind doing your filming from one spot.

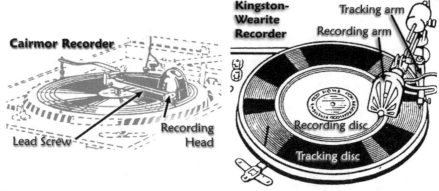

Fig 8-2. Domestic audio recorders from the 1930s. The Cairmor recorder-adaptor (left) and the Kingston-Wearite recorder (right).

From 'Amateur Talking Pictures', B. Brown, 1933

In late 1930, a device went on sale for attaching to a conventional domestic gramophone. Manufactured by Cairns & Morrison Ltd of London, the *Cairmor* converted the gramophone into an audio recorder (see Figure 8-2). The price included six 7-inch diameter aluminium blank discs labelled *Silvatone*. The Cairmor was only one of many such systems each offering slightly different methods of recording. They all were as simple as the Edison Phonograph, though using soft aluminium as the recording medium. The track pitch was determined on the Cairmor by a lead-screw mechanism for traversing the cutting head across the disc surface. There were two other systems that used lead-screw tracking: the 'Ecko Radiocorder' and the 'Harlie' Recording Attachment.

The lead-screw made it tricky for the cutting head to follow the natural undulations of the aluminium blank. Another approach, used in the 'Kingston-Wearite' recorder, was to have a 'tracking disc' (see Figure 8-2). This had a space in the middle for carrying the blank. To the outside of the disc surface was an unrecorded cut grove. The record cutting head was on a conventional arm, with a spur attachment holding a stylus. This stylus followed the outer groove, guiding the cutter across the blank's surface. In the United States, home audio recording was far more developed and had a far greater market than in Britain. Products such as the Pam-O-Graph and the Trugraph were popular, high-quality self-contained audio recorders.[7]

Vision Disc Transcription

I had transcribed the Phonovision discs in the early 1980s using domestic-grade equipment, suitably modified for Phonovision. The master transfers were made onto half-track professional analogue reel-to-reel magnetic tape at 15 inches per second – the best available format at the time. In 1996, the National Museum of Photography Film and Television at Bradford decided to have archival quality digital audio transcriptions of all the Phonovision discs. Knowing the locations of the discs and the important issues surrounding transcribing them, I welcomed an invitation to be involved in the transcription project.

In preparation for the transcription, I met with Eliot Levin, Managing Director of Symposium Records and the man who would make the archival transfers for the Museum. The lessons learned from Phonovision had proven helpful and allowed the transcription to go smoothly. Although the frequency correction for gramophone discs is normally the RIAA curve (a standard still in use today in audio amplifiers for those still using vinyl records), the restoration work suggested better results would be achieved with a flat response – the 'Blumlein' characteristic. In addition, all audio-specific effects for cleaning-up the signal were bypassed. The transfers were the closest to the raw signal on the discs. During transcription, the video images were monitored in real-time on my software version of a Televisor display to ensure the best transcription quality.

It turned out that the new transcriptions made by Levin were equivalent to the transcriptions I had made over a decade before. There were to be no new revelations.

Revelations from Aluminium

In the course of discussing the idiosyncrasies of having video recorded onto an audio disc, Levin happened to mention he had a friend, Dave Mason, who owned a home recorded aluminium audio disc with what might be a television signal recorded on it. The disc had a standard Silvatone label with 'Television 1933' written on it in ink (see Figure 8-3). The only way I could be sure of what it contained was to look at the signal on my computer. Levin generously agreed to do a quick transfer for me to check out.

Fig 8-3. The Silvatone Souvenir aluminium disc holding the earliest known recording of broadcast television.

Original photograph courtesy of the NMPFT
From the collection of David Mason

When the tape copy arrived, it became clear that it held something unique and unlike anything that I had seen before. This was a true 30-line signal heavily distorted by the home-recording process and years of corrosion of the aluminium. Also unusual was that there was a great deal of movement on the television picture. There were snatches of recognisable images, but not enough to be certain.

One part of the recording stood out clearer than the rest. I could just make out a troop of dancing girls doing high kicks. This was completely stunning. After 13 years of having only the Phonovision discs and the Major Radiovision disc, and a lifetime of being brought up to believe that broadcast 30-line television was amateurish and staid with stilted performances, this material seemed completely alien. A call to Levin and a few days later, Mason's disc had been expertly transcribed to DAT. Transcribing the disc was quite a challenge for Levin. Over its surface, there were great patches of aluminium oxide corrosion. Not only had his stylus to plough through these damaged areas, it also had to track grooves that shallowed to almost nothing.

Pictures from the first BBC Television Service

For quite a while, I was convinced that this was not a studio performance, but part of a movie scanned into the 30-line picture. That would have explained the massive amount of movement, the vitality of the performers and the professionalism of the production. As the restoration progressed, it became obvious that the performers were limiting their actions to the 30-line frame. There were no cuts between scenes and the camera panned across the dancing girls from side to side in a technique called 'hose-piping' – a true sign of there being only one camera. This was undoubtedly a recording of a studio performance made, if the label was correct, in 1933.

Two clues to dating the disc came from studying the television picture. First, there was none of the geometric distortion caused by a Nipkow disc camera. This meant that the camera was based on a mirror-drum. Second, the camera panned across the subjects, but did not tilt when the dancers came close in. This was exactly the feature of the BBC's single 30-line mirror-drum camera in use from the start of the service in 1932 right through to the end of the 30-line service in 1935.

The preliminary restoration process took several days – considerably longer than for a Phonovision disc. This was not surprising, as, regardless of the far lower quality, the Silvatone disc comprised around 2,700 frames compared with a Phonovision disc's 700.

Over the following few weeks, the Silvatone disc succumbed to more software image processing, almost all of which had to be written from scratch to deal with the unique problems. Unlike the Phonovision discs, the image suffered from complex phase errors, high surface noise, drop-outs and complete gaps where playback had failed. There was a serious 'Catch-22' problem: the timing correction software required the drop-outs to be corrected and the software for cleaning up the drop-outs required the timing to be correct.

After processing, the quality of the four-minute silent video recording was adequate to recognise what was happening. The dancing girls were fascinating subjects. They were wearing dark one-piece bathing suit outfits and were bareheaded. Although I was curious about what this programme could be, the likelihood of finding which of the 1,500 possible programmes this was seemed remote.

One Sunday morning whilst thumbing through my small collection of early television magazines, looking for material for my website, I was stopped in my tracks by a caption-less picture (see Figure 8-4). In this picture six girls, wearing what looked like one-piece dark bathing costumes to my fashion-less eye, were clustered around and draped over a grand piano. The article was a review of the television highlights of 1933. I immediately rang Ray Herbert, who also had the same magazine, and told him the page number. He came straight back with the answer – the Paramount Astoria Girls. The next port of call was the BBC Written Archives Centre at Caversham, near Reading. Their search for the Paramount Astoria Girls in 1933 yielded only two possibilities: the Paramount Astoria Girls appearing in April, and the Paramount Victoria Girls in August. Comparing the 'P-as-B' (Programme as Broadcast) listings, the only match was for the April programme. In an outstanding piece of luck, this programme could only be the revue 'Looking In', which was broadcast on the night of 21st April 1933.

Fig 8-4. The Paramount Astoria Girls in rehearsal with Pepper, Watt and Robb for the 'Looking In' programme. April 1933

From 'Television' magazine, January 1934

After the Daventry National Programme finished at 11:10pm on the 21st April 1933, the following took place, lasting from 11:12pm through until 11:53pm:

Television Transmission by the Baird Process
(Vision 262.6m; Sound 398.9m)

"LOOKING IN" specially written by John Watt
Music by Harry S. Pepper
Produced by Eustace Robb in conjunction with the Author and Composer
Cast:
Iris Kirkwhite

Anona Winn

Veronica Brady

Horace Percival

The Paramount Astoria Girls

Small Orchestra:
Doris Arnold (pianoforte), J Hanrahan (Drums), S. Kneale-Keley (Violin), J. Romano (Saxophone)

Checking through the reports in the magazines brought out something even more remarkable: the 'Looking In' programme was advertised as a historic television first – the world's first ever television revue. Incredibly, the picture I had found by accident was taken at the rehearsal for the programme on the Silvatone disc. There were a few other pictures of the Girls in rehearsal for this programme, showing them doing their high-kicking dance routine.

It would seem that an enthusiastic amateur saw that there was a television special being broadcast and set up his recording equipment to 'have a go' at capturing it. Judging from the state of the disc and ignoring the effects of ageing, he would have had difficulty seeing much more than the odd glimpse of the girls doing their dance routine. In a strange way, this is probably the longest time-shifted video recording ever, lasting 63 years. Far more important is that this video recording gives us our first-ever look at exactly what people were watching on television in the early 1930s.

This is a landmark reference recording – the first discovered recording of a broadcast television programme – and one that easily challenges a half-century old myth of poor quality programmes and an amateurish service.[8]

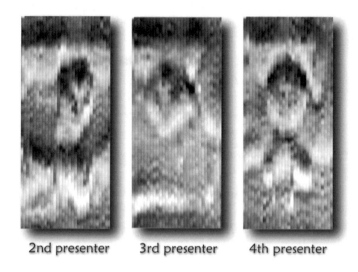

2nd presenter   3rd presenter   4th presenter

Fig 8-5. Single frame images from 'Looking In' depicting three of the presenters, as they appeared one after the other. Each presenter moved with radically different gestures. Though not obvious from the still, the 4th presenter (right) has both hands holding his lapels. Despite considerable restoration, these images are still poor quality.

Courtesy of the Author

'Looking In'

Despite its poor quality, this amateur recording is arguably more important than Baird's experimental Phonovision recordings made professionally over five years before. It lets us see through the eyes of the viewing public just eight months after the start of the BBC 30-line Television Service. What they saw has come as a great surprise: fast-paced entertainment, full of movement.

At the start of the four-minute long recording, an insignia on a curtain (for all intents a caption) is brought forward towards the camera in a zooming action with the focus being pulled to track the movement. The camera of course was unable to zoom, since such a lens had yet to appear. One after the other, five presenters stand up (rather than enter from the left or right) into shot from behind the curtain. This curtain hangs down to the level of the performers' lower-back just at the bottom of the field of view. Each performer either talks or sings with plenty of movement and then ducks down out of shot. Once they have finished, some 96 seconds into the recording, seven small women appear one after the other. We know they are small because they appear in the lower half of the picture. (The camera could tilt up and down, but only with some difficulty.) These seven women are the six dancers, plus one of them possibly coming back for a second

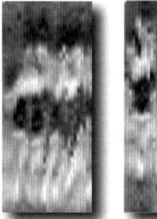

Fig 8-6. The Paramount Astoria Girls performing their high-kicking routine in long-shot. Still images do not do the performance justice, as the movement brings the image alive. However we can make out two dancers on the left and almost three on the right.

Courtesy of the Author

appearance. Each dancer pops her head into view, one after the other, from underneath the curtain. (The contemporary reports of the programme describe the performers bursting through paper hoops. If they do, they are not visible on this recording.) The curtain is lifted out of the way, the camera rolls focus back to a long shot and we see the six dancers in one-piece bathing suits entering stage right and performing an 80 second dance routine (see Figure 8-6).

This routine is synchronised high-kicking (see Figure 8-8), similar to that of the Tiller Girls who regularly appeared on 'Sunday Night at the London Palladium' in the 1960s. We can tell how fast the music was played from the girls themselves. The interval between the girls' kicks (averaged over ten full 'kick-cycles') is 760 milliseconds, giving a tempo of 79 beats per minute.

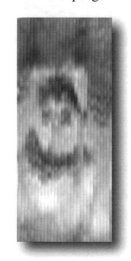

Fig 8-7. This presenter appears after the dance sequence.

Courtesy of the Author

At the end of the sequence, the girls dance off to stage right, still kicking. After 8 seconds of blank screen, we see a close-up of an announcer with a squared-off headdress (see Figure 8-7). After a further few seconds, we reach the end of the disc.

The 11 or 12 performers and presenters appear one after the other within the space of 2 minutes. Not one of them is in shot for more than 20 seconds. This is long enough to be recognised but not so long that the viewer gets bored.

Fig 8-8. The Paramount Astoria Girls in 1933.
Courtesy of the Author from original courtesy of the BBC

Showing the dancers in close-up first, then straight afterwards in long shot was a technique mastered by the producer, Eustace Robb, that had been described, but never before been seen. Tony Bridgewater, who mixed the sound on these 30-line programmes, independently described Robb's technique in interview in 1984.

> 'Eustace Robb opened our eyes to what could be done. Take the ballet for example … He adopted all sorts of clever techniques (to make the most of having only 30-lines). First of all the dancer would come into close-up so that you established what she looked like and who she was … and then, very gradually, she'd trip backwards and you wound the focus on the camera and followed her and began to see more and more of her. And then, when she got far enough back against the backcloth, she could actually go into her dance routine. And you could even get two or three figures doing this.' [9]

Notably, on the dance sequence in 'Looking In', the long-shot view of the girls dancing shows around two to three girls in the frame at a time (see Figure 8-6).

Movement – The Key to Clarity

The computer-processed stills and the off-screen time exposures in this book only hint at what people really saw. Judging the 30-line television system by the quality of these stills would be a mistake. Though we have only 30 lines and only so much detail along a line, we are missing a crucial element: time.

Fig 8-9. The 30-line Television Control Room, 16th March 1934. The camera/projector is at far right. At lower left is a caption scanner, with a Nipkow disc flying spot projecting onto a board. Two angled vertical boxes hold small photocells in what is a miniature of the main studio. By 1935, the simple caption holder was replaced by an octagonal rotary unit, allowing rapid change of caption.

Courtesy of the BBC, RTS37-85

Complex images, such as a picture of a back garden, just do not work in 30-lines, whereas a person moving, such as a gymnast, comes across incredibly clearly. If we freeze the scene and just look at one static 30-line picture, the image becomes far less clear. As the image is then a still built up from just 30 lines, it is difficult to distinguish between what is image and what is not. A single frame of the Paramount Astoria Girls may be crudely recognisable, but when seen as a moving dynamic television image, the Girls come to life before our eyes. There is something strange at work and it has much more to do with what we perceive than what is there in pixels, lines and frames.

What we are experiencing is not the detail that the eye sees, but the recognition of movement that the brain sees. For movement that we would instantly recognise, our brain somehow builds up a model of what we are looking at. We interpret the model into a real-world scene.

Low definition television becomes more effective, the more movement there is. Eustace Robb understood the limitations of the 30-line system and the benefit of movement. He offered the viewer in the disc's four minutes plenty of variety and movement that would not go amiss in a modern TV commercial advertisement. His effort captured on disc is even more impressive when we appreciate the basic nature of the equipment he used.

Today many TV producers would balk at the technical limitations under which Robb worked (see Figure 8-9). From the small sample that we have of Robb's work, we would be hard-pressed to find a producer who could create as interesting a programme geared and tuned to the features and limitations of the 30-line system. Imagine making a half-hour entertainment programme with your own camcorder. It has to be fixed to the floor on a tripod such that you can only pan the camera from side to side. You cannot zoom or record and edit the programme (the technology had not been developed in the 1930s). With only one camera, you cannot cut from one shot to another. Not only that, but the programme you are making is going out live to an audience of several thousand across Britain. Your video controls are the lighting and the camera's focus. You have one advantage over the BBC in the early 1930s; your performers are not working in a pitch-black studio with an arc light flickering in their faces.

Reviews of the Revue

The reviews were extensive – much more so than for other programmes. The BBC had staged what they claimed was the world's first television revue. Its importance was underlined as the books on the BBC Television's history by Ross and Swift both covered this programme. By good fortune, we seemed to have discovered four minutes from a historic and well-documented BBC programme. However, this is just the sort of programme that would have attracted interest and therefore the most likely to be recorded by anyone reading about the forthcoming programme in the Radio Times and the Press.

The review in 'Television' describes the programme in some detail.

> 'Just when I was settling down to assess the month's efforts, thinking that all the big excitement was over (possibly referring

to the start of BBC Television some 8 months before), along comes the first television revue to upset all calculations... Production was in the hands of Eustace Robb, whose experience in Studio BB (in Broadcasting House) made it possible to present this ambitious effort. Six Paramount Astoria Girls were picked for the show from a team of twelve, trained by Mrs Rodney Hudson. They all wanted to come along, and selection was difficult. Colouring was the decisive factor, blondes do not show well ... *(in a long-shot)* and brunettes were chosen. John Watt had an idea to show them doing physical drill in white swimming suits *(that would have been impressive – the white suits would be the same colour as flesh, rendering them naked!)* but Robb ruled this out ... so we saw girls in black in their opening number.' [10]

Viewers (or 'lookers-in') from across Britain saw details that are not evident on the disc recording. One person in Newcastle showed that his preference matched mine over sixty years later. 'In my opinion, the outstanding features of the entertainment were the dances given by the Paramount Astoria Girls...' [11] From London, where reception was better, one 'looker' described the revue as 'a remarkable new achievement for television which should surprise the critics ... The inclusion of the Paramount Astoria Girls in their programme was ambitious but eminently successful...' He closed with this comment on the future of television: '...sound without sight lacks kick.' [12]

In the reviews following the performance, the Press gave varied responses, but generally recognised this programme as being a step forward in television.

The Daily Telegraph: 'While demonstrating the remarkable technical progress made in transmission, the reception of this programme clearly showed that stage performances cannot yet be regarded as an ideal subject for television.'

The Evening News: 'Television as an entertainment has taken a step forward. The whole performance was vivid and lifelike ... There is no doubt that television as a form of entertainment has come to stay.'

The Daily Herald: 'The first revue to be televised, with all the faults of mechanism still needing perfection, was more exciting than the wireless (programme) which immediately preceded it. Stage plays want sight as well as hearing. Television moves on, slowly but surely.'

The Games Discs

Two years after the Silvatone recording was discovered, a pile of home recordings on aluminium discs was found in a house clearance. Jon Weller, a collector of old electronics equipment, retrieved the collection. Marcus

Games, a keen amateur movie enthusiast, had recorded them supposedly in the 1930s. Weller later discovered that several of the discs had unusual material recorded on them (see Figure 8-10) and contacted me through the Narrow Bandwidth Television Association (NBTVA).

Fig 8-10. One of the 'Marcus Games' discs. This one contains a fragment featuring Betty Bolton and is marked rather impolitely 'Woman Large Head'.

Courtesy of the Author

The soft aluminium recordings once again dictated special treatment though less in Eliot Levin's transcription than in the computer processing. Though these discs were also domestically recorded aluminium discs, the faults on the recording made them as different as chalk and cheese to the Silvatone disc. The physical quality of the recording was superior to the Silvatone disc with far less of the corrosion problem that hounded its transcription. Each disc held one or two fragments of 30-line television recorded at various speeds from 100 to 120 revolutions per minute. On each of the discs, the recording speed had drifted steadily throughout the recording. In addition, the size of the signal, equivalent to the deviation in the groove, progressively dropped throughout the recording. Unfortunately, this was not enough to identify the recorder. However, as the faults differ from the Silvatone 'Looking In' recording, we can at least tell that the recorder was not the Cairmor.

Of all the 30-line disc recordings, the discs of Marcus Games are by far the best quality. Regrettably, Games did not leave behind any document-ation on the recordings. There was a description written on one of the discs, but no dates and neither indication nor proof that the description was written at the time of recording. The restored recordings – eleven in total – appeared to be a collection of fragments of solo singers and some unusual material that suggested a children's programme.

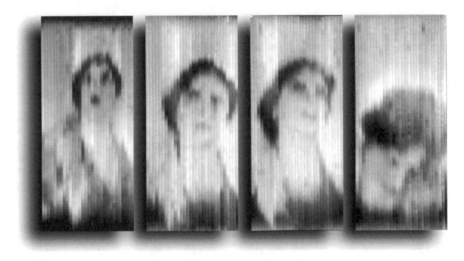

Fig 8-11. Stills from three separate disc recordings show the same performer, whose gestures are reminiscent of a classical or even operatic singer. Between mid-shot (left) and close-up (middle), she goes 'off-stage' rather than moving closer. The still on the right is her head bowed at the completion of the song.

Courtesy of the Author

Authentic or Fake?

The plain aluminium discs hold deceptively high quality images. Without any dates on them, the discs could have portrayed 30-line television from almost any time. Could these discs have been made more recently?

The only other source of low definition television in Britain is the Narrow Bandwidth Television Association (NBTVA). They have adopted, for amateur television use, a standard of 32-line television with an aspect ratio of 3 vertical to 2 horizontal. They have also elected to include embedded electronic sync pulses that go below black level, borrowed from contemporary television.

The signal on the discs is simply nothing like the NBTVA standard. The aluminium discs sport a Baird broadcast format of 30-lines of what looks like 7:3 aspect ratio. As on the 'Looking In' recording, the lack of geometric distortion means that a mirror-drum camera was used. Not only that, there is no discernible geometric error caused by misplaced apertures, strongly indicating the camera was precision-manufactured and professionally constructed.

One of the most striking and distinctive features is the lighting. Unlike any of the valiant efforts of the NBTVA, the scene was professionally lit. In the case of the male operatic singer, the lighting is so good that we can see details on his dark jacket and on the brightest parts of his face.

Assuming the conventional 30-line format for scanning was in use, the performers all exit stage left (to the right as we see it). This is consistent with the BBC studios in both Broadcasting House and at Portland Place.

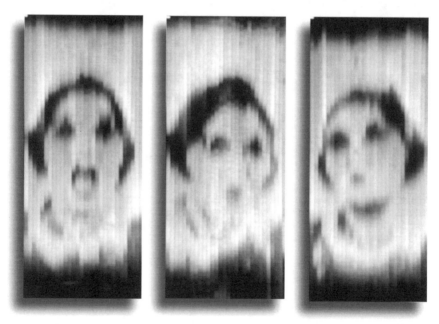

Fig 8-12. Single frames from the recording of Betty Bolton. The make-up on the side of the nose is less pronounced than that of Jane Carr (Figure 7-13), suggesting a date later than 1932.

Courtesy of the Author

Immediately, stage right, there was a heavy curtain used to block off the lighting from the orchestra or band from the darkened studio. Of course if this series of discs were off-air recordings of BBC programmes – the most likely explanation – then the studio 'lighting' would have been done by positioning photocells, as described earlier.

The camera was able to roll focus between head-and-shoulders close-ups to mid-long shots. In the background, we can see the back of the studio at more than 6 metres away from the camera. This was a big studio and well lit all the way to the back. On top of that, the camera was able to follow the action by panning to keep the subject perfectly in frame. This precluded any amateur offering, though the closest possibility were the Holmes brothers from East London. Their televising area was not a studio but a portrait area about the size of an A4 sheet of paper.

The singers were all highly professional (see Figure 8-11), none more so than a female with straight black hair (see Figure 8-12). Now it is of course possible, given there is no evidence of recording date, that these are recent recordings of performers in period costume. However, when we consider the professionalism of the engineering of the camera and the almost perfect lighting, such a reconstruction seems extremely unlikely. The evidence very much supports these as being genuine off-air video recordings of 30-line television broadcasts from the BBC in London.

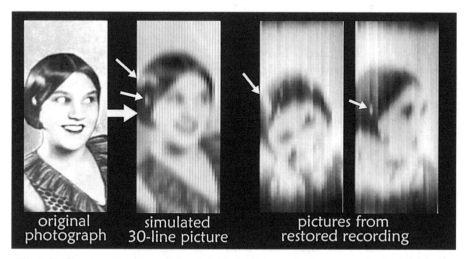

Fig 8-13. Visual confirmation of Miss Bolton's features came from converting a 1929 image of her (left) to a simulated 30-line image and comparing the features – such as the high reflection from the hair and the silver clasp holding the kiss curls in place.

Courtesy of the Author

Betty Bolton

The lady with the straight black glossy hair and the kiss curls, held in place by silver clasps (see Figure 8-13) gives an outstanding performance, superior to that of the other singers. This is Betty Bolton. Though the features on the recording match those of Miss Bolton's (see Figure 8-14), confirmation came when I took the restored video to her daughter, Judyth Knight. At that time, I had an open mind as to who the singer on the disc could be. After all, I had never seen Miss Bolton perform and only knew how she looked from stills. Initially, Miss Knight had difficulty with the image on the screen; the 30-line format was new to her. Miss Knight remarked that the singer's hair seemed to be thicker on the top of her head than in the photographs. What she was seeing was the effect of phase error, which appeared to extend her hair upwards slightly from the top of her head. Once she got used to seeing the 30-line image, she became convinced it was her mother. Not only was the profile perfect for her, Miss Knight became certain when she recognised her mother's *mannerisms*. This was a genuine recording of Betty Bolton.

Fig 8-14. Studio pose of Betty Bolton.

Courtesy of Betty Bolton

Betty Bolton's Career

Between 1930 and 1935, Betty Bolton appeared on 30-line television some thirty times, eighteen of which were on Baird Television from the Long Acre studios. Miss Bolton told me that she had a flat nearby in the West End of London, making it easy for her to drop by after giving a performance in one of the many London shows. She was already an accomplished singer and dancer by the time of her television appearances.

Fig 8-15. Portrait of Betty Bolton, the first performer on the BBC 30-line Television Service.

Courtesy of Betty Bolton

She had started out in vaudeville at the age of ten, giving such an impressive performance that she featured on a full-page spread in 'The Sketch'.[13] She entered the world of show business and dance bands. She appeared in several films, notably 'Balaclava' and 'Wolves'. She was a radio star through her frequent appearances on BBC Children's Hour, as assistant to Kenneth McCulloch. By the late 1920s, she was singing with several of the more popular dance bands in London. The 1930s were a busy time for her not only in television but also in the shows of the West End of London.

Betty Bolton was the first performer to appear on the BBC's 30-line Television Service on Monday 22nd August 1932, singing and dancing in the new television studio in the basement of Broadcasting House. She went on to give a further twelve performances on the 30-line service.

After getting married in 1936, domestic pressures meant that she had to curtail her career. As a result, this star of vaudeville, theatre, dance bands, radio, records, the movies and television, dropped out of circulation.

In 1992, the BBC celebrated 70 years of radio and 60 years since the start of the BBC 30-line Television Service. In an attempt to find her to celebrate her contribution, Tony Bridgewater re-discovered her by placing a personal advertisement in the Daily Telegraph. Her re-discovery in 1998 from the 30-line recordings sparked media interest yet again. She featured on the new digital widescreen format, looking at her restored 30-line television image silently singing on the 65-year-old recording. Recorded on 13th June 1998, this was subsequently broadcast on the inaugural programme of the BBC's digital television service, echoing her inaugural BBC TV appearance all of 66 years before.

Dating Miss Bolton

BBC's 'Tomorrow's World' was interested in the latest discovery and recorded a programme as a follow-up to the discovery of the 'Looking In' recording.[14] One of the researchers approached a lip-reader used by the BBC. The quality of the 30-line image was simply not good enough. Although the vowel shapes show clearly, the lip-reader needed to see the lips to read them. Miss Bolton told me that, from her actions, she believed that she had been singing a love song. Her popularity on 30-line television precluded finding out in which of the programmes she featured. Singing several songs for each performance and giving a few dozen performances throughout the 30-line period created too many options to track down the programme.

The best guess is that this was a programme transmitted on the BBC

30-line service sometime between 1932 and 1935. With the large line-blanking area on the video signal and the high quality of image, even from the back of the studio, it is highly likely that these recordings were made from transmissions from the upgraded studio at BBC Portland Place, effectively dating the recordings to between 1934 and 1935.

The other performers – male and female operatic style singers – give performances that are more stilted but appropriate to an operatic style of singing. One recording captures the last few seconds of a female singer who blows several kisses before disappearing. She does not blow them *at* the camera, but off to one side – so that we can see them clearly (see Figure 8-16).

Fig 8-16. An unknown female popular singer (left) and an unknown operatic male singer. The middle picture of the girl shows her blowing a kiss obliquely to the camera (to make it visible). Even from these still frames, the detail on the man's jacket, collar and tie show up well. (right)

Courtesy of the Author

The recording of the male operatic singer (see Figure 8-16) shows lighting that would be unusual today. When he moves his arms around, we see a shadow cast upwards across his jacket. The lighting engineer, D. R. Campbell, had placed a photocell bank just below the camera view-port. More than likely, this was to catch detail from the singer's jacket. The result works well, with his lapels and top breast pocket showing quite clearly. In surviving photographs of BBC Portland Place, we can see such a photocell on a short stand.

Gramovision

The Televisor manufacturers lobbied for greater airtime for 30-line television at more convenient times of day. They wanted to be able to demonstrate television to customers rather than sell inert boxes. By 1934, it became obvious that the service was destined for closing. Nevertheless, at least one entrepreneur thought it would be valuable to have television available at any time. As we read earlier, Barton Chapple had declared in 1934 that it was feasible to 'bottle' or 'can' television using domestic audio recorders. Whether inspired by this, or by earlier articles on Phonovision, announcements soon appeared on the availability of pre-recorded 30-line videodiscs.

> '...a London firm is making them for initial distribution to radio dealers for demonstration to the public. These records ... have a double track for sound and vision and are played with a special electrical pick-up which passes the signals to the vision and sound reproducers respectively. It is understood that these first records are made from the 30-line television programmes provided by the BBC, if they do prove satisfactory, experimenters will welcome them as a means for providing test signals at any time of the day instead of reliance being placed on the two television transmission periods now available each week.'[15]

In 1935, an advertisement appeared for a 30-line videodisc for the purpose of testing and lining up displays. The manufacturer was the Major Radiovision Company, whose General Manager was F. Plew. The disc was to be sold for seven shillings, with sixpence for postage and packing.

The distinctive red label of the Major Radiovision disc graces many a private audio collection today. This is the only surviving recording, of which there are many pressings, from the promised pre-recorded 30-line

24-HOUR TELEVISION

Moving vision frequencies have been successfully recorded on a gramophone disc. Simply connect your Pick-up to your Television Receiver. One stage of amplification is sufficient.

Plew Television Records give you Moving Television whenever you wish. They are double-sided, run for 6 minutes each side and have the life of a normal record.

Now that the B.B.C. Television transmissions have temporarily ceased, Plew Television Records are invaluable to all who have a Television receiver.

LIMITED QUANTITY AVAILABLE AT PRESENT

Send to-day for Record No. 2 (with Synch. Signal). Gives Twelve Minutes of Television as often as you like.

PRICE **10/-** Post Free

F. PLEW (122), 70 MARYLEBONE LANE, LONDON, W.1

Plew Television Records have been bought by The Royal Military College, Camberley, and many other Technical Colleges. Now essential for Television Demonstrations.

Fig 8-17. Advertisement for Plew's 30-line video test records. Plew was also General Manager of 'Major Radiovision Company', which had earlier sold one of the two commercial discs that survive today.

From 'Practical Television' magazine, November 1935

videodiscs of the mid-1930s. The other recordings such as dual video/audio recording identified as 'Gramovision' have not survived and indeed may never have appeared as products.

For every Phonovision disc found, there were ten Major Radiovision discs. Major Radiovision was recognisable as a 10-inch (25 cm) diameter double-sided 78 rpm disc with what looks like track banding across its surface with 10 'tracks' per side. The red label sports 'Recorded Television Record Speed 78, Scanning speed 750 (rpm), Lines 30', with an unusual 'For Private Use Only' underlined. This conventionally recorded 78 rpm disc was in no way synchronised to the scanning mechanism and was intended to be played back on a conventional record deck. The electrical audio signal had to be connected to the amplifying section of a broadcast radio and the Televisor connected as for normal 30-line broadcast reception. Instructions on

Fig 8-18. The Major Radiovision disc was a 30-line video-only test disc that was sold for seven shillings by mail order and through Selfridges department store in London.

Courtesy of Author

how to connect the output of a gramophone playing this disc back appeared in a magazine.[16] However, a description of what 'lookers-in' would have seen never appeared.

Major Radiovision – A Major Disappointment

The material on this disc is distinctly uninspiring. Each side consists of a series of still pictures that we see being slid into position, left in place for twenty seconds or so and then slid out to be replaced by the next one. Without true movement, this disc is in no way representative of 30-line broadcasts. Two of the stills are in reality the same picture, with one the mirror-reversal of the other. It looks like the pictures are transparencies, or magic lantern-slides, that have been lit from the back. The mirror reversed slide was the same transparency simply slid in the wrong way round. Looking closely at the pictures, we can see classic signs of arc-scanning.

The camera was a Nipkow disc. However, in 1934, when this disc was on sale, Nipkow discs had long since been replaced by mirror-drums.

The purist would argue that this really does not qualify as television. The recording could have been made using a strong light shining through the transparency and focused onto the surface of a conventional Nipkow disc from a Televisor. A photocell behind the disc would gather the light and convert it to the signal we see on the disc. This is more like 'telecine' or the early shadowgraph experiments of Jenkins and Baird from years before.

The disc is supposed to be a collection of high-quality 30-line TV test pictures (see Figure 8-19). There it fails miserably. The quality suffers from a 5 kHz 'ringing' that may have been a physical resonance of the head-cutter. Only after using filters to cut out the resonance (and consequently the high frequencies of the recording), does the high quality of the images become apparent (see Figure 8-21).

We saw earlier that centring the disc is essential to get a stable playback. As the disc was to be played back on a conventional audio turntable, there is only the centre spindle hole to locate it. This is nowhere near what is needed to generate a stable picture on a Televisor. Even with the disc perfectly centred and the turntable quartz-locked for speed, we see today that the recording itself was not made under such good conditions. The image drifts and bounces. As a TV test signal disc, it is a total failure. Without movement on the disc, it also fails to bring across the capabilities of the 30-line medium.

Today this disc is touted as a Baird disc and has even been described as Phonovision. Phonovision it most definitely is not. The disc is only

Fig 8-19. The still pictures that comprise the double-sided Major Radiovision 30-line video test disc.
Courtesy of the Author

'Baird' in that the signal recorded on it complies with the 30-line format. It is doubtful that Baird, the Baird Company or even the BBC had anything to do with this disc.

A 'Sister' Disc to Major Radiovision

One other 30-line recording exists, but only as a poor transcription on audiotape. Unfortunately any information on where and what the original disc was, has been lost. Like the Major Radiovision disc, this disc also comprises a sequence of lantern-slides. Each of them is slid into place, left for a time and slid out in a manner indistinguishable from the Major Radiovision recording. There is no 5 kHz resonance, giving a clear view of what is on the disc. Despite the clarity of the recording, the lantern slides are of quite unrecognisable shapes (see Figure 8-20). This may possibly be Plew's 'Number 2' disc advertised in 1935 (see Figure 8-17). With only a tape copy and no disc or documentation, we can go no further.

Fig 8-20. A still from another 30-line test disc.
Courtesy of the Author

The BBC's 30-line Recording

Amongst all the recordings in the BBC's extensive sound archives, there is only one disc of a recording of 30-line television signal. This has throughout the years caused confusion with people finding one of the pressings and believing they have come across an undiscovered recording. The problem stems from the official printed label. It reads, 'Sample of 30 line Baird Television Transmission (copy of a recording of an actual transmission)'. The double-sided 12-inch (30 cm) 78 rpm disc is identical in content to the common Major Radiovision disc. The quality of the BBC disc is much poorer than the Major Radiovision disc. This suggests it is a transcription, rather than a parallel off-air recording. The poor quality is surprising but probably caused by handling the original signal as if it were an audio recording rather than a television signal.

Fig 8-21. A still from the Major Radiovision disc.
Courtesy of the Author

For all we know, the Major Radiovision disc could have been transmitted despite all the faults listed above, but there is no documentation to prove this one way or the other. It is safer to assume that the label is simply wrong and that the BBC disc is therefore just a copy of the commercial Major Radiovision disc.

Changing History?

Those 'lookers-in' of the 1930s have captured far more than they could ever have imagined. They have inadvertently left us not just the earliest, and consequently the longest, time-shifted videos in history, but a historical reference about what the first television audiences really watched. In the case of the 'Looking In' recording, one person has deliberately captured the first few minutes of what was considered a landmark television programme.

Audio gramophone discs remained only an undeveloped possibility for recording 30-line television. Most certainly, despite the efforts of Baird, it did not become the first *practical* television recording method. It is through using a computer that some of the faults have been corrected, allowing us to see those silent ghostly images from the dawn of television entertaining us once again.

The tendency in computing today is to reconstruct and simulate what may have been seen. This is especially important in archaeology where the past is being reconstructed with mere scraps of evidence. Whilst important for visualising what may have been, simulation techniques applied to the 30-line imagery are beyond what is appropriate. Careful restoration, rather than interpretation, has been the keynote to the work described in this book. I leave it for others to use their imaginations and their ingenuity to recreate those events.

---

[1] SAGALL, S.: 'Television in 1934', *Television*, Jan 1934, p6

[2] BENN, J.: 'Time is Running Out', *Television*, May 1933, p187

[3] KAYE, R. C.: 'Making Television Records', *Television*, Feb 1932, pp470–471

[4] PALMER, F. G. R.: 'Gramophone Recording', *Television*, Aug 1932, p225

[5] BARTON CHAPPLE, H. J.: 'Canned Television', Draft article for *Practical Television*, 3rd Mar 1934

[6] BARTON CHAPPLE, H. J.: 'Canned Television', *Practical Television*, Nov 1934, pp55–56

[7] BROWN, B.: 'Amateur Talking Pictures' (Pitman), 1933, pp29–43

[8] WARD, M.: 'Pioneering TV girls dance again', *New Scientist*, **152**, No 2053, 26th Oct 1996, p22

[9] NORMAN, B.: 'Here's Looking at You' (BBC/Royal Television Society), 1984, p83

[10] ANON.: 'Last Month's Programmes', *Television*, p170, May 1933

[11] ANON.: 'Reports from Readers', *Television*, p171, May 1933

[12] ANON.: 'Reports from Readers', *Television*, p171, May 1933

[13] ANON.: 'Child and Super-child, Betty of the Vaudeville', *The Sketch*, 16th Aug 1916, p141

[14] *BBC Tomorrow's World*, 9th Sep, 1998

[15] ANON.: 'Gramovision', *Practical Television*, May 1935, p216

[16] ANON.: 'Television from Gramophone Discs', *Practical Television*, June 1935, pp257–258

9 Capturing the Vision

"The video hardware market is a graveyard for technology."

Anon. c1985

Television - the Ephemeral Medium

Today, we switch on our TV sets, select our favourite channel and sit down to watch ... videotape. Television has become largely scheduled replays of videotaped material. This is especially true for most of the cable and satellite TV channels. The major networks, such as the BBC and ITV in Britain, NBC, ABC and CBS in the United States, provide, as part of their integrated offering, a live news service and live sports and events coverage. However, this service inevitably includes videotaped items. Even CNN, the epitome of instant round-the-world news, relies heavily on videotape. That is the inevitable outcome of our increasing demands for a high quality service.

The original objective for television was to 'see at a distance' *at that instant*. Its live nature was television's edge on the motion picture industry. We could sit at home and watch with everybody else a monarch being crowned in Westminster Abbey, a horse crossing the finishing line at the Derby, a man making his first steps on the moon. It could also allow us to be in a place where the situation would normally preclude us: a sole protester blocking the passage of a tank in a city square or pictures from the front-line of war in a desert. Aided by satellite communications, television allowed us to watch revellers around the world as the year number changed from 1999 to 2000. Time-zone after time-zone, people in every country seemed to engage in a planet-wide Mexican wave to celebrate, amongst other things, the decimal system of counting.

When taken with video recording technology, television still provides all of the above and much more besides. Recording allows us to focus on just the highlights, capture the events for posterity, and compile and edit raw material to create an item of far greater interest than the original. In that sense, making a television drama today is similar to making a motion picture. However, we sometimes need reminding that the movies led the way.

We benefit today from the best features of television: instant communications for timely events and a programmed schedule for high quality entertainment. Above all, we have the ability to instantly switch to live coverage when the unexpected occurs.

Recorded television is without doubt central to all of these services today. Though the primary reason for recording is to time-shift the event to be more convenient to broadcast network schedules, it also allows us to keep the best of that material for later. Archiving the major events gives us the benefit of seeing programmes and events of the past. All those archived television recordings do more than simply entertain. They give us a remarkable insight into how we used to live, who entertained us, what was going on in the world, and even how the overall message of television was presented. We can even use such an archive to study the more sinister uses of television, through its control and manipulation for personal, corporate or political gains.

How far back?

With the exception of the gramophone video recordings, existing video material does not go back to the beginnings of television itself. Any search for early TV material starts running out before the middle of the 1950s. There is almost nothing from before the late 1940s – over ten years after high definition electronic television started and twenty years after the very first television programmes. The reason is simple: the technology for video recording is far more recent than for television itself. Video recordings on tape appeared from the early 1950s and film recordings of a television image just after the Second World War. Those define how early we can go.

What confuses the picture is that we in Britain do at first glance appear to have material that goes right back to the beginning of high definition television broadcasting, to a time before the Second World War. These television-related items are however not television recordings but either films that became inserts into programmes, films made at the same time as the television transmissions, or films staged and re-shot as 'television'. Occasionally we get a few snatches of off-screen television recorded on film for newsreel and documentary, though they are incidental to the film.

A great example of re-staging a television event for film appears in the now historic documentary 'Television Comes to London'. It covered the start of television broadcasting on 405-lines. Whenever documentary-makers want to use early British TV material, they invariably reach for this film as a primary source of stock footage – and usually for either Helen MacKay singing 'Here's Looking at You', or Adele Dixon singing 'Magic Rays of Light'. The performances are very close to what people would have seen, but they were filmed separately for the benefit of the documentary. The new service started in November 1936, stopped for the War and re-started in June 1946 with the same faces and almost the same equipment.

The earliest known occasion of a deliberate recording of BBC 405-line television was in 1938 – just two years after the service started. Andy Emmerson, a British TV journalist, discovered this recording in 1999.[1] The four-minute silent film contains snatches of 405-line BBC transmissions apparently from 1938. The quality was poor, but then the film was recorded at Schenectady, New York in the USA, from a direct VHF transmission from Alexandra Palace in London, England – a distance of over 5,300 km. The result is a heavily distorted picture, with tantalising snatches of scenes from forgotten programmes, including some glimpses of Jasmine Bligh reading out announcements. Unfortunately, the film gives no real clues about what the material was or when exactly the programme was captured (see Figure 9-1).

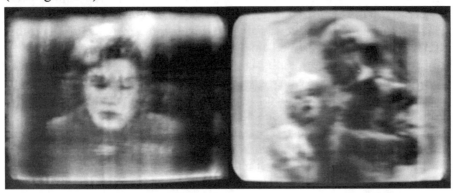

Fig 9-1. Stills from the 1938 off-screen recording made in Schenectady, NY of BBC transmissions from London. On the left is Jasmine Bligh.
Courtesy of A. Emmerson, enhanced by the Author

Pre-War British Video Recording

For the entire broadcast period of 30-line television – for Baird's experimental TV programmes from 1929 and for the BBC's first Television Service from 1932 to 1935 – all the directly recorded video material we have is contained on the gramophone videodiscs described in this book. Other than those discs, there is nothing.

For the 30-line material, both by Baird and the BBC, the displayed image was simply too dim to be captured onto film. In the early days, the display system was a neon lamp behind a Nipkow disc of pinholes. The faint neon glow from any one part of the image appeared for the instant the aperture was over that spot and it was a full TV frame time later (80 milliseconds) that light again came from that spot. An exposure in the order of several minutes was quite normal to capture a single image. Consequently in Britain at least, off-screen photographs from this period are rare and movie films non-existent.

The Early International Scene

The global picture of early recordings of television is very sketchy. The tremendous achievements in Russia, Hungary, Germany and Japan do not appear to have included a method for recording the television signal. That achievement, if it can be so called, seems uniquely British. However, there is no good reason why it is unique to Britain. Anyone could have attempted to record a television signal onto an audio disc, provided the signal was of low-enough definition.

In the United States, C. Francis Jenkins was the primary experimenter in mechanically scanned television. As far as video recordings were concerned, there is no evidence that he ever attempted to make them.[2] Throughout the USA, there is only one documented instance of recordings of US mechanically scanned images. These, however, were of facsimile still images. For some months in 1928–29, the station WOR transmitted facsimile images on the Cooley Ray-Foto system. Austin Cooley and Edgar Felix, editor of Radio Broadcast magazine, recorded the vision signal on an Edison wax cylinder Ediphone for the convenience of playing back the signal at any time to test the equipment.[3] The feeling is that the only amateurs to record television onto cylinders or discs were those that got the idea from Baird's example, well publicised in British journals. In the USA, there were no similar publications and consequently no known attempts at recording US television images.

Of the main US corporate organisations, the Radio Corporation of America, RCA, was the most influential in US television history. From its company acquisitions in 1929, it rapidly swept the mechanical systems to one side to make way for electronic systems. The man behind RCA's rise to success was David Sarnoff, President of RCA from January 1930 and recently described as the 'godfather' of television.[4] Alex Magoun, Curator for the Sarnoff Corporation, told me the issues that steered RCA away from the mechanical path and hence from the possibility of any early videodisc recordings:[5]

> 'First, once Sarnoff acquired the services of Vladimir Kosma
> Zworykin and the engineers from Westinghouse and GE in

1929–30, the mechanical tradition … rapidly faded under Sarnoff's and Zworykin's determination to televise electronically.'[6]

'Second, there was some determination that the quality of service had to be substantial to justify more than marketing for the sake of novelty. RCA experimented with field-sequential technologies in 1931–32 and was planning to introduce some form of TV in 1932, but the Depression (and the quality of the system) at the least stopped the Company.'

'Third, the acquisition (by RCA) of the Victor Talking Machine in 1929 led to tensions in RCA over the relative importance of electromechanical versus purely electronic products. This came to head with the VideoDisc (in the 1970s) and I can tell you that there are still two sides to that technology dispute among the RCA veterans today.'[7]

The Bell Telephone Labs succeeded in recording vision by a much more successful method than the Baird wax discs. Dr Frank Gray, co-author with Mertz of the definitive theoretical paper on scanning, announced in February 1929 that they had developed a method of directly recording onto 35 mm film.[8,9] In effect, Gray and his team had built the first *tele-recording* equipment. The image was 50-lines at 18 pictures (TV frames) per second, far higher in information content than would be possible to capture faithfully onto wax discs.

The following year, the General Electric 'House of Magic' labs at Schenectady, NY, recorded television in a similar fashion onto film. This could be seen as the forerunner of tele-recording (see Figure 9-2).

Fig 9-2. Continuous frames from a recording by General Electric Company's 'House of Magic' at Schenectady, NY. *From Cameron, 'Radio Television', 193*

Recording Methods

Development in electronics gave us high definition 405-line television broadcasting in Britain by 1936 and 525-line television broadcasting in the United States by 1941. The new electronic technology left all the available recording methods far behind. Video recording would not catch up with those other developments for almost 20 years, until the 1950s. Home video recording would have to wait a further 20 years beyond that, until the 1970s (ignoring some of the failed formats, such as the 1960's Telcan and Wesgrove linear video tape recorders).

None of the very early recording systems for television had been perfected for faithful and consistent reproduction. They did however give images of sorts, but not of a quality that broadcasters could find acceptable. Those broadcasters increasingly demanded a method of recording that gave on playback an image indistinguishable from the original broadcast.

Prior to the Second World War, there was no capability to capture and store the live television image faithfully for re-broadcast. Television was operating as a real-time live system. As television began to relay major news events, the first being the 1937 Coronation, there came the need to create some archive of the event, other than by sending newsreel cameras to film alongside the television cameras.

Tele-recording

After the Second World War, television companies developed a practical means for making film copies of televised material from a precision TV display. The challenge with electronic television was to get the TV display correctly lined up for exposure and dynamic range and to have perfect synchronisation between film camera and TV display.

Quite simply, if the camera filming the screen was not running at exactly the same rate and precisely in step with the displayed picture, then the picture would pulse in brightness. We can see this effect when we point a TV camera or camcorder at a computer monitor tube (rather than an LCD (Liquid Crystal Display) screen). The different picture refresh rates give a horrendous flicker at the beat frequency or difference in camera and display rates.

In Britain, the process of making precision film copies of television was called television film recording or just tele-recording. In the United States, the same process was called kinescope recording. The quality of the film recording was difficult to maintain. Even with synchronisation of the film and the video signal, tele-recording always showed up some other visual problem. The geometry of the set-up or the alignment and long-term drift of the picture's brightness and contrast often caused heavy distortion.

The primary role of tele-recording was to create a copy of the programme that could be used for distribution and for time-shifting rather than as a programme production tool. Despite the introduction of tele-recording in broadcast operations in the late 1940s, television was almost exclusively live until the late 1950s. The BBC broadcast all public events

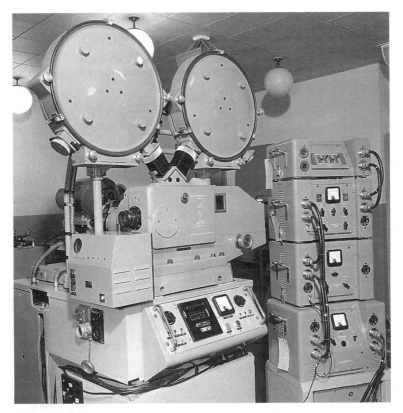

Fig 9-3. The Moy-Cintel 35 mm television recording camera, Serial No. 2. Manufactured for the BBC by Ernest F. Moy Ltd of London

Courtesy of the BBC, RTS37-77

live, such as the Victory Parade in June 1946 and two major events of 1948: the Olympic Games and Royal Wedding of Princess Elizabeth and Philip Mountbatten.

The Oxford–Cambridge boat race of 1949 was one of the first events to be tele-recorded off-screen onto film. For a time tele-recording was the only method of capturing and archiving television material – allowing us today to share in some of those historic events (see Figure 9-3). For the historian in media studies, the kinescope recordings of American shows of the late 1940s and early 1950s are a goldmine of early material. Looking decidedly amateurish and low budget by today's standards, these programmes were all performed and transmitted live. They have a homespun character all of their own and are a reminder that, without the Heath Robinson approach to kinescope or tele-recording using film, we would not have this material today.

The mediocre quality and practical difficulties of tele-recording led to serious effort to solving the problem of recording video. The image just had to be captured at source, in the native electrical video format, and some method had to be worked out to do it. With no prospect of inventing a completely new process of recording, some means of adapting what was already possible seemed to be the only answer.

At the start of tele-recording, there were only two viable technologies capable of recording a signal waveform (excluding optical soundtrack recording). There was the gramophone (needle-in-the-groove) mechanical recording process and the magnetic recording process onto tape. Neither of these were particularly new methods. Both had been available at the time of the 30-line video recordings.

These technologies lie at the root of the development of video recording. The story of their development puts video recording into context with recording technology generally.

Fig 9-4. The Edison 'Triumph' Phonograph Model A, introduced 1900. This model is serial number 37839, circa 1903, modified with 4' gear, introduced in 1908.

Courtesy of the Author
From the collection of E. B. Levin

Recording – from the Beginning

The home audio revolution is not recent; it started in the 19th century with the first methods of recording using the phonographic principle. The method was very simple: a surface coated in a soft material, such as wax, was turned underneath a tiny cutting stylus. The first machines used cylinders mounted on rotating spindles rather than flat discs – the Phonograph (see Figure 9-4). The outside surface of the cylinder was covered with a material that could be marked (like tinfoil or wax) and a stylus attached to a diaphragm traversed the surface. Sound from the performer caused the diaphragm to vibrate. The stylus traced out a gentle vibration in its track across the soft recording surface (see Figure 9-5). On replay, the stylus followed those gentle undulations, producing a faint replica of the original sound through an acoustic horn loudspeaker.

Fig 9-5. A 4-inch (100 mm) recorded brown cylinder showing a modulation pattern (from a fault during recording) similar to what we would expect from a cylinder recording of Baird's Phonovision.

Courtesy of the Author
From the collection of E. B. Levin

Edison made the Phonograph practical in 1877 when he demonstrated the first successful recording and playback of a human voice. His tinfoil-covered cylinder captured the nursery rhyme 'Mary had a little lamb...' and the world was changed. We take recording so much for granted today that it is difficult to understand the impact it had on people's imagination back when it was first demonstrated. The editor of Scientific American captured the mood at the time:[10]

'It has been said that Science is never sensational; that it is intellectual, not emotional; but certainly nothing that can be conceived would be more likely to create the profoundest of sensations, to arouse the liveliest of human emotions, than once more to hear the familiar voices of the dead. Yet Science now announces that this is possible, and can be done ... Speech has become, as it were, immortal.'

Edison was not the first with the idea, he was just the first both to make it work and to promote it successfully. Léon Scott de Martinville had developed a write-only system based on marking a pattern on a soot-coated paper wrapped around a cylinder. The 'Phonautograph' was sold as a scientific instrument for studying the patterns that sound made. The mechanical arrangement was in principle the same as Edison's. The difference was that Scott's invention came out *twenty* years before Edison's – in 1857.

Though cylinders and their players were easy to manufacture, it was quite a different matter to mass-produce pre-recorded material. The problem with replicating cylinder recordings was that several had to be recorded at once in real-time. Emile Berliner came up with the perfect answer for pre-recorded material; record them on the surface of flat discs. Berliner's discs could be replicated by stamping them out from a master, making copies instantly and more importantly, rapidly and at low cost (see Figure 9-6).[11]

Launched in 1893, Berliner's 'Gramophone' and discs struggled against the cylinder, eventually prevailing after a messy period of lawsuits. The cylinder lingered on in the niche area of office Dictaphones but could not seriously compete with the mass-produced disc. The gramophone went into the language and Berliner's disc started an entire industry that lasted from the 19th century into the 21st century.

The beauty of the cylinder and early acoustic gramophone disc recordings is that electronics were not essential, only being needed to amplify the signal during recording and playback. For many years, all recordings were done acoustically – using merely the energy in the sound waves and sophisticated shapes for both microphone and loudspeaker as acoustical amplifiers.

The 1920s saw the emergence of electrical recording. Its development required thinking about the complete recording chain – from microphone right through to pressing – rather than focusing on any one single component. As with domestic hi-fi today, the results were only as good as the weakest part of the system. The focus was on establishing exactly how to improve all of the elements in a balanced way: microphone, amplifier, disc cutter, wax disc itself, the manufacturing reproduction process, the pick-up and the playback unit. By 1925, Western Electric offered vastly improved quality of recordings, though still on 78 rpm discs. This was the technology in use at the Columbia Graphophone Company at the time of Baird's Phonovision recordings.

By 1950, the biggest change in the gramophone disc had come with new plastics for the pressing material. The shellac 78 rpm disc had ruled for decades as the primary medium for carrying audio recordings to the buying public. Its popularity meant that the '78' continued to be sold well into the 1950s long after the new vinyl discs had gone on the market. The vinyl recordings could be made longer with a quality that was vastly superior to the '78'. It took a few years for the first discs to appear, and to be adopted as a replacement format by the public.

Fig 9-6. An example of a 7-inch diameter Berliner Gramophone disc, dated 13th March 1899.

Courtesy of the Author
From the collection of E. B. Levin

CBS brought out the $33^{1}/_{3}$ rpm (revolutions per minute) long-playing (LP) discs on vinyl in 1948. The LP and the 45 rpm 'single' (released in 1949) dominated the home audio entertainment market for nearly 30 years.

The Compact Disc took over from LPs from the 1980s onwards, yet vinyl lingered on in the 12-inch (30 cm) 45 rpm dance music 'single'. In early 2000, the Compact Disc format was already showing its age when compared with the advanced functionality of MiniDisc and the derivatives of DVD.

Magnetic Recording

1900 was the year that Perskyi introduced the world to the word 'television'. Elsewhere in Paris that same year, at the Exposition Universelle, a Dane called Valdemar Poulsen was celebrating winning the Grand Prix du Paris. He was exhibiting his invention – a recording machine that used magnetism on metal wire instead of the cut of a needle in a groove (see Figure 9-7). It was not fully perfected and gave poorer results than cylinder and disc recording technology. It however started an industry in wire recorders and led, after many years, directly to the tape recorder.

Wire recorders did have a sizeable market because of the ability to re-record many times. Their poor quality kept them out of the home and in use for telephone answering machines and Dictaphones.

Fig 9-7. Valdemar Poulsen's Telegraphon magnetic wire recorder.
Courtesy of the Science Museum, London

With the advent of electronics, it became practical to make magnetic recorders. One of the first machines to be put into broadcast use was the *Blattnerphone*. The BBC used these unwieldy monsters from 1931 until the Second World War – indeed Neville Chamberlain's declaration of war on Germany in 1939 was recorded using a Blattnerphone.[12] The Ludwig Blattner Picture Corporation of London had joined forces in 1929 with the German company, Telegraphie-Syndicat to produce the machine. Intended for recording soundtracks for the motion picture industry, it was used extensively by the BBC for its radio programmes. Its benefit was in sounding (over the medium waveband) almost as good as the live performance. Shellac disc recordings had a distinctive surface noise and scratches that destroyed the illusion of being 'live'.

The Blattnerphone used solid steel tape wound on huge reels and passing the head assemblies at 1.5 metres/second. The tape was thin and sharp, and at the speed it travelled, it was potentially dangerous. There are no recorded amputations arising from accidents, but the machine gained a great deal of respect. When editing the tape, there was the delicate job of welding the two parts together and keeping your fingers crossed (the ones you had left) when the join ran past the recording head.

The Beginnings of Magnetic Tape

The Germans mastered the practicalities and materials science for magnetic tape recording during the Second World War. Although the Allies understood the principal of magnetic tape recording, the Germans had developed it to such an extent that the German hardware was used for many years after the War.

In 1944, a Signals Corp technician, John T. (Jack) Mullin, was part of a team scavenging any electronic items of interest left behind by the retreating German army. He found and sent two working Magnetophone tape recorders and around 50 tapes back to the USA. On 16th May 1946, Mullin was giving a presentation and demonstration of the German equipment at a meeting of the Institute of Radio Engineers in San Francisco. The quality of the audio and the ease of using tape (duration of recording and simple, clean editing) captured the imagination of Bing Crosby Enterprises – the company that managed and looked after its namesake. The problem they had was that Crosby wanted some time off, yet radio demanded live performances of him during prime time. He needed to sound live – something that disc recording just could not handle. Mullin, the two German recorders and the 50 German tapes were used to record the Bing Crosby Show for a season.[13]

By April 1948, the first US-made magnetic tapes appeared, to be played back on Ampex tape recorders – initially direct copies of Mullin's German Magnetophone machines. In Britain, EMI developed the BTR1 – another copy of the German Magnetophone. So successful was the BTR1 that it became the mainstay of professional audio recorders until the 1960s.

Despite television being around since the 1920s, the technology to record it for broadcast use had to wait on the development of tape systems. Only by putting the development of the audio tape recorder in such a context do we now see why the video recorder did not appear before the 1950s.

Video: faster and higher

The top frequency that could be recorded and replayed for all the differing recording formats was limited by the technology. For gramophone recording, the higher the frequency, the more rapid the cutter had to move. The upper limit was set by the recording medium and the drive electronics. Gramophone recording was constrained to being an audio format. The mechanical action of ploughing the signature of the sound into the warm wax during recording effectively limited the top frequencies to be within the audio band.

Magnetic recording though had different constraints. With carefully designed recording and playback heads, the top frequency for tape was determined largely by the speed of the tape across the heads: the faster the tape moved, the higher the frequency recorded. Getting a sufficiently high speed for tape travelling past the pick-up and recording heads was going to be the key to video recording.

In the early 1950s, the highest frequency in the US system was around 5.5 MHz and in the British 405-line system, around 2.5 MHz. At that time, there was no hope of mechanically or optically (as on film) recording such a video signal. This left magnetic tape as the *only* medium available.

The late 1940s and early 1950s saw not just a push for professional audio recorders but for a practical video recorder. It seemed that it would be feasible to record video onto tape so long as there was a high enough speed for tape moving past fixed heads. That simple fact focused research on both sides of the Atlantic.

The video signal for television presents a major challenge to tape recording technology. For audio, the ideal recorder has to capture all the frequencies and dynamics that the ear can detect: from very subtle *pianissimo* sounds, to extremely loud *fortissimo* sounds and from very low frequencies (around 20 Hz) to the limit of human hearing (around 20 kHz) or around 10 octaves. For video, the recorder has to be able to capture a far greater frequency range – of around 18 octaves – and the recorded signals comprise timing, brightness and colour information, usually along with audio.

The real challenge is in the massive range of frequencies to be captured onto magnetic tape. Fundamental to tape recording is that the signal output from tape is a function of frequency. The signal output rises at 6 dB per octave. This means that for 18 octaves, the electronics around the record and replay processes has to deal with 108 dB of dynamic range. This is the single biggest problem facing designers of videotape recorders and caused the first such systems to split the video signal across multiple channels.

In 1951, John Mullin was working with the Crosby organisation in developing a video tape recorder – VTR – from an extension of the audio tape recorder. Only eleven months after starting, the team received encouragement with a first crude demonstration in November 1951. They managed to get the tape speed down to 100 ips (inches per second) by spreading the video signal across 10 parallel tracks, with two other tracks for timing and sound. A year later and the quality had improved considerably.

The race was on and RCA brought out their own VTR, still pulling tape rapidly across a stack of fixed heads. RCA's VTR captured full-colour TV using five separate linear tracks spread across normal half-inch magnetic tape. The video signals for the three component colours, Red, Green and Blue, occupied three of the tracks. The other two tracks, like Crosby's VTR, were used for timing and sound. The tape speed was an alarming 360 ips (about 32 km/h) and a 7,000 ft (2.1 km) reel lasted only four minutes. It seems unusual today, but one of the USA's major networks – CBS – almost gave RCA an order for machines. The order was not made because the Ampex Corporation, working in secret for several years, lifted the covers off their radical new design, and changed television forever.

Ampex

In 1956, thirty years after the first demonstration of television by John Logie Baird, and twenty years after the start of high definition electron tube television broadcasting, Ampex announced the world's first truly practical video recorder. The problem had been how to move the tape past the recording heads at the speeds necessary for video. The RCA, Crosby and later the BBC approaches were to have stationary heads and move the tape at high speed – a linear video recorder. This meant these solutions all had mammoth tape reels and 'battleship engineering' to move kilometres of tape in just a few minutes at a precisely controlled speed. Consequently the programmes were extremely short and media costs were high. Ampex turned the problem round. They moved the heads past the tape.

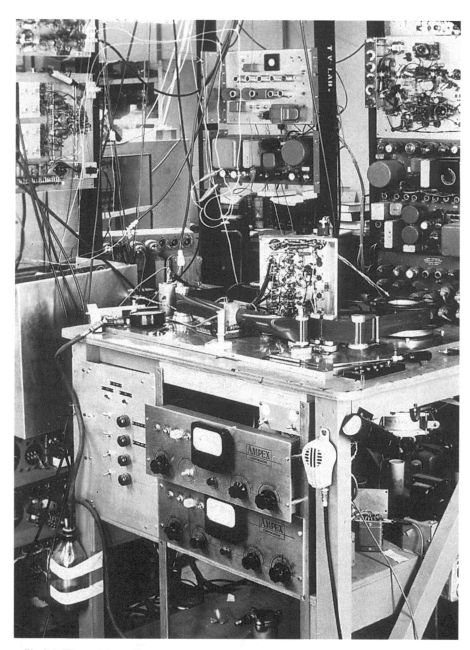

Fig 9-8. The prototype Ampex rotary head video tape recorder, VRX1000.

Courtesy of the Ampex Corporation

In their production systems from 1956 onwards, they used wide 2 inch (50 mm) tape, mounted four heads equally around a disc and spun the disc at high speed across the width of tape. The four-head arrangement gave rise to the name for this type of recorder – the 'Quadruplex' (otherwise known as 'Quad', 'two-inch' or 'segmented'). The tape was moved in the usual direction but it only needed to move slowly – at 15 ips, the standard speed for audiotape. The movement caused each pass of a head across the tape to create successive parallel *swipes*. The wide flexible tape had to be 'cupped' into an arc so that the heads on the edge of the disc would evenly swipe across the tape. Each swipe was long enough to allow the end of one swipe to be duplicated at the start of the next swipe – avoiding breaks in the signal. Mechanical scanning for television made a comeback … in a manner of speaking.

The Development of the Practical Videotape Recorder

Ampex managed to develop, build and manufacture a world-beating machine of unique design (see Figure 9-8). The success story started when Ampex senior engineers met with Camras of the Armour Research Foundation.

One of the most prolific inventors in the field of magnetic recording was Dr Marvin Camras. Born in 1916, his work on developing wire-recorders led him to a career in the Armour Research Foundation (later the IIT Research Institute) that lasted from 1940 through to 1987. In that time, Camras collected more than 500 patents through developing new techniques in multi-track recording, magnetic sound for the movies and videotape recorders. He received many honours, culminating in the National Medal of Technology awarded by President George Bush in 1990, five years before he died.

As a recognised expert in the field of magnetic recording, Camras was consulted by Ampex to establish a method of using magnetic tape for recording a video signal. He came up with a technique using rotating heads, an idea that had been around since the 1930s, when it first appeared in the German 'Tonschreiber'. In December 1951, Alexander M. Poniatoff (the AMP of Ampex) authorised the development of a video recording machine using rotating heads.

Scanning the video heads across the tape at high speed meant that the signal would have to be segmented or broken up into swipes (see Figure 9-9). Though this technique would achieve the required high head-to-tape speed, there would also be the problem of a break in the signal as one swipe finished and the next was being written. In addition, quite unlike linear recording, there had to be a means of somehow synchronising the tracks on playback. The playback heads had to read precisely along the centre-line of the track.

To minimise the effect of breaks in the signal, the swipe length ideally ought to have corresponded to one complete picture. This would have allowed the heads to change over between the end of one picture and the start of the next, out of the viewable area of the picture. However there were distinct problems going down that path, enough for the Ampex engineers to try an alternative approach.

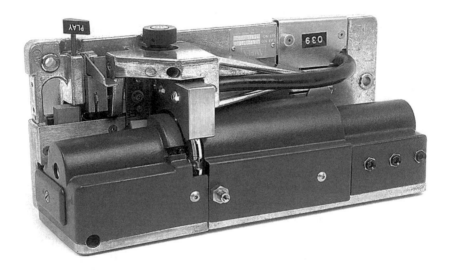

Fig 9-9. The head assembly from an early Ampex Quadruplex.

Courtesy of the NMPFT

Quadruplex versus Helical-scanning

The head-to-tape speed for broadcast quality pictures was around 1,600 ips (40.6 m/sec) in the Quadruplex videotape recorder (see Figure 9-10). Using that tape speed, the length of tape for one complete picture or field would therefore have been around 32 inches (81 cm). With 2-inch (5.1 cm) wide tape, this meant that the heads would have had to swipe across the tape at a precise but shallow angle. The heads were mounted in a cylinder whose circumference had to be precisely the length of tape for one picture. The tape was wrapped around the cylinder at a slight angle, giving rise to the appearance of a helical path to the tape. This gave rise to this recording method being called 'helical-scan'.

One of the challenges of using helical-scan for broadcast use was the sheer scale of result – dictated largely by those early low-density tape formulations for video. The precision necessary for having a fixed head follow a path less than 0.3 millimetre (0.01 inches) wide for two-thirds of a metre, and do this reliably from machine to machine on flexible, almost fluid, magnetic tape was challenging. The primary solution that eventually made this practical was many years away. It entailed mounting the replay heads on piezoelectric or voice-coil transducers and shifting their position dynamically to follow the subtle changes in the track. This was what was needed for a broadcast standard helical-scan machine.

In the course of their development of the Quadruplex machine, the Ampex engineers understood the benefits of helical-scan recording. They however did not believe they could create a solution acceptable for broadcast use. Consequently, they pursued the transverse scanning approach using short segments across the width of 2-inch (5.1 cm) tape. They trail-blazed new techniques to handle the needs of the videotape recorder and the idiosyncrasies of their segmented scanning approach.

The single most important step in making videotape recording practical was in the development of F.M. (Frequency Modulation) for recording.[14]

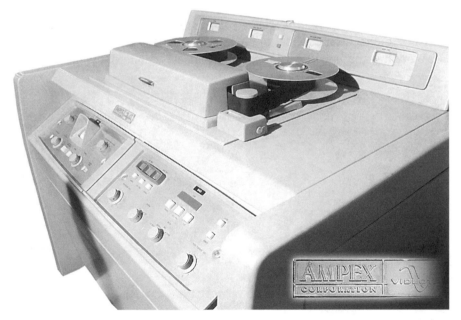

Fig 9-10. The tape unit of an Ampex VR1000B.

Courtesy of the Author
From the collection of the NMPFT

Though putting the video signal on a higher frequency carrier reduced the number of octaves, using the signal to vary the frequency of the carrier rather than its amplitude (A.M.) brought major improvements to the quality of recorded video.

After an intensive but successful development period with some of the best engineers in the industry, Ampex launched the Quadruplex on an unsuspecting broadcast television community on Saturday, 14th April 1956. They demonstrated two complete machines at two locations: one at the National Association of Radio and Television Broadcasters convention in Chicago, Illinois, and the other at Ampex's Redwood City, California headquarters. Charles Ginsburg, the team leader, described the response.

> 'The demonstrations were a bombshell in the industry. In Redwood City the performance was sensational, exciting and satisfying. In Chicago, pandemonium broke loose and Ampex was flooded with orders ... From that beginning, videotape recording grew into a billion dollar industry that touches our lives in some way every day.' [15]

Helical Scanning Takes Off

Toshiba of Japan, rather than Ampex, developed and pioneered the helical-scan videotape recorder.[16] Without any previous experience of magnetic recording, Toshiba began work in 1954 (see Figure 9-11) and introduced their first monochrome model to the public in 1959. They had succeeded by downgrading the specification from broadcast standard by dropping the

Fig 9-11. The prototype hand-made Toshiba helical-scan machine – the VTR0 – used one rotating head in the canted cylinder left of centre. Pictures of sorts were recorded and replayed in 1956, supporting the case for further development.
Courtesy of the Toshiba Corporation

head-to-tape speed substantially, thus making the machine practical to build. Toshiba's demonstration at the SMPTE (Society of Motion Picture and Television Engineers) convention in 1960 showed the benefit of making the swipe the same length as one complete picture. By changing the linear tape speed, they could demonstrate slow motion, still-frame or reverse-motion.

Right through to the 1970s the helical-scan videotape recorder was the mainstay of CCTV – Closed Circuit Television. It was used in industry and commerce and in private television studios such as, for example, in Universities. With improvements in magnetic tape and developments in electronics, the helical format eventually ousted the Quadruplex and became the format of choice from broadcaster to homeowner.

VERA Lingers

At the time when Toshiba in Japan had first succeeded in recording and replaying vision signals from a helical-scan machine and when Ampex in the USA were demonstrating their Quadruplex machine in 1956, the BBC Research Department was several years into developing its own video recorder. This was known as VERA (Vision Electronic Recording Apparatus). Unlike the US and Japanese major developments, the BBC had opted to continue development of a linear video tape recorder (see Figure 9-12).

Since Britain's monochrome 405-line TV system had far less information in it than the USA's NTSC colour 525-line system, the BBC engineers only needed to move their tape at 200 ips (18 km/h). Even so, recordings lasted just 15 minutes. With the Ampex Quadruplex machines entering service in November 1956 with CBS, and later in 1957 with NBC and ABC, the BBC continued development of VERA. They gave their first on-air demonstration in the summer of 1958 and the plug was pulled on VERA shortly afterwards. In hindsight, and in comparison with the previous US commercial developments of Crosby Enterprises, RCA, Ampex and others, the BBC's VERA stands out as being unusual – almost an academic exercise. VERA occupies a place in Britain's television history that substantially overstates its importance.

In 1958, a few months after VERA's first demonstration, the first two Ampex Quadruplex machines not only in Great Britain but Europe (the VR1000, specially modified for the 405-line UK standard) went into service with Associated Redifusion in Kingsway, London.[17] A year later and the number of Quadruplex machines had grown to four and then to sixteen by 1962. Initially these machines were used in a similar fashion to tele-recording, solely for time-shifting completed programmes.

Fig 9-12. One of the BBC's VERA linear video recorders in 1958.
Courtesy of the BBC, RTS37-96

Their primary defined objective in the USA was to provide a delay in transmitting programmes to accommodate the US time zones. On this basis, original estimates predicted a world-market of around seven to ten video recorders. Their use as a production tool, through tape-editing programmes directly on videotape, was actively discouraged initially. The perception was that it would tie up valuable and expensive machine-time and *increase* production costs as a result. This situation was soon reversed and the videotape recorder became an integral part of programme making.

Tape – the Fluid Medium and the Store House

Many think that recording a television signal onto tape is an engineering marvel. The creativity and inventiveness in getting practical working systems is the marvel, but the process in reality is a 'kludge'. If we had some better and cheaper means of capturing and storing video, we would use it. Magnetic tape is a poor medium for archiving and needs elaborate handling of the television signal before it can even begin to do the job of

storing the signal. Once it is stored, there is the problem of getting the signal back off the tape. Here even more elaborate processing is needed to take out timing jiggles, to correct for tape drop-outs and to maintain a good quality replay of the signal. The tape itself is a fluid and flexible medium and all the elaborate hardware and processing is there to minimise the effect of its deficiencies (see Figure 9-13).

So why do we use it? Tape is the least worst solution to storing television, yet is also the most cost-effective storage medium in use today. Tape can be thought of as a *volume* storage device. Although the surface of the tape is two-dimensional, the tape is

Fig 9-13. A Sony 0.5 inch head assembly. On the lower half of the drum, a tape guide rises from left to right. The edge of the tape would follow this ensuring the heads sweep across the tape at precisely the desired angle.

Courtesy of the NMPFT

wrapped onto a spool, forming what could loosely be considered as three-dimensional recording.

All other storage methods – needle in a groove, images on film, computer discs (hard and floppy), CD or Laserdisc – use a two-dimensional surface to capture the data. DVD is one tiny step beyond, as it comprises two storage layers. The computer hard disc goes beyond that in having a stack of multiple discs operating in parallel.

Strictly speaking, the technology for all of these is two-dimensional storage. We would gain so much in capacity and hence costs associated with storage if we could only use the *volume* as storage. Its absence is not for want of trying. Many of the world's corporate research laboratories are trying to get to the 'sugar-cube' memory. This would be a device having a three dimensional structure of memory elements, where each element would be accessible by its x, y and z address – rather than something that dissolves in a hot drink.

One of the problems of a reel of videotape is accessing anything within this volume. We have to physically 'spool' the tape in fast-forward and fast-rewind – and that takes time. The worst video recorders for locating a scene on tape were the linear video recorders. Their 'Play' speed was also their top winding speed. Consequently, they did not have 'Fast Forward'.

Trying to find anything on the tape accurately can be difficult. Recording a time-code sequence with the video and audio is the professional answer, but the problem remains for other analogue formats. We have all come across this when we try to find 'that song' on an audiocassette tape.

Although we would prefer something simpler and cheaper to use, there is no alternative to magnetic tape for the near future. In television, tape has proven successful as a means to time-shift and repeat programmes and live events, and as a medium for editing and creating programmes. That we do not know whether we are watching a live programme or a recording is a measure of how successful videotape recording is today.

Disc Technology

Discs are a much more convenient format than tape. They provide access to any part of the recording very rapidly and are handier to use and to store. Baird had used discs to record Phonovision simply because the gramophone disc was still the only suitable format available to the public. The home audiotape recorder was over twenty years away.

Being able to access a scene rapidly became the primary reason for using discs in broadcast television. The first such system was the Ampex HS-100, released as a product in March 1967. It could do slow motion and stop-action replays and was a boon for televising sports events. With two large discs, providing four recording surfaces covered in a magnetic coating, the system could buffer over 30 seconds of broadcast quality video. The discs were synchronised to the television signal and were spun so that they had one TV field (half a TV frame) per revolution. The heads tracked radially allowing a still picture to be played back repeatedly by simply keeping the head in one position. Moving the heads slowly gave slow motion replay.

Emile Berliner believed that discs were superior to cylinders because they were easier to mass-produce. For the same reason, the manufacturers of domestic electronics believed that videodiscs would be superior to videotape. Visit any videocassette duplication house and you will see racks of video recorders; each machine making their own real-time copy of the master quality recording, in a similar fashion to audio cylinder production in the 1900s. (High-speed 'contact' duplicators for videotape have not been as commercially successful as their audiocassette counterparts.) Though not exactly using softened shellac under a heavy steel press, making multiple copies of videodiscs should be less time consuming, cheaper and therefore more profitable. This approach spawned a range of domestic formats. All but one of these failed soon after their release.

One of those failed formats was the TelDec, latterly TeD. A joint venture by Telefunken in Germany and Decca in Britain, the company produced a videodisc playback system that bore some resemblance to Baird's 1927 Phonovision. First announced in 1970, the system was developed over the next five years to achieve colour playback with stereo sound from a thin floppy PVC disc (see Figure 9-14). It was only capable of around ten minutes continuous play, limiting its usefulness to music videos and the like. It used a needle-in-the-groove approach that caused

Fig 9-14. The TelDec player and flexible videodisc removed from its protective cover. The market concept of this device – able to play short video programmes – was similar to Baird's Phonovisor of some 50 years before.

From originals courtesy of the NMPFT

appreciable wear. The quality of playback was not good – using a line-sequential approach (red line, followed by a green, then a blue) to achieving colour. Though roughly a third of the price of a domestic videocassette recorder at the time, the lack of suitable material and quality effectively killed off the format.

The successful videodisc type – the Laserdisc – thrived only in the United States, surviving elsewhere as a specialist format for those seeking near perfect video quality. First introduced in 1978 in the USA as Laservision, it struggled to be adopted in Britain (see Figure 9-15); just succeeding in the early 1990s with a revision to its specification that incorporated digital audio.

As soon as DVD was released in the late 1990s, there was a rapid adoption of the new format. Players capable of both CD and DVD playback became cheaper than dedicated CD players, became 'big-sellers' in supermarkets, and the discs themselves hit the market only slightly more expensive than audio CDs. DVD looks to be the format of choice for pre-recorded video material for some years to come.

Fig 9-15. A European Philips Laserdisc player, VLP700.

Courtesy of C. D. McLean

There is no time like the Present

Of all television images, we can say with some certainty that those of the first steps on the moon, on Apollo 11 in July 1969, will be preserved for as long as humankind exists. In that sense, the Apollo 11 imagery is the extreme opposite of the 30-line recordings. At least there is a recognised need to provide permanence for such historic events.

However, even the 'forgotten' can turn out to be amongst the most important imagery. The first images restored from Baird's Phonovision discs elicited this response from Granada TV production staff, '(the material) has had an impact at least equal to footage of the Moon Landing. Gasps of admiration were heard from all present'.[18]

The restored 30-line recordings have passed the first hurdle of preservation: their historical value is now recognised. Like all other recordings – including the most precious, they now face the challenge of the 'arrow of time', what they say in physics as 'increasing entropy'.

Who wants to live forever?

No medium of storing images today can survive forever. That is absolutely fundamental. Even more important is the understanding that the methods of recording today – optical, mechanical and magnetic and all their variations – have a finite life measured only in decades. Of these recording methods, magnetic tape is the worst and needle-in-the-groove mechanical is the best.

Television has for the most part been recorded on magnetic tape since at least the 1960s. The longevity of videotape is already a major problem. With the development of all the different methods of recording, far more serious though is the problem of finding machines to play back archive material.

The Explosion of Formats

The first practical helical-scan video recorder from Toshiba was the progenitor of all analogue video recording formats. Since then, there have been a great many variations on the helical-scanning theme. Just as an example, the *main* domestic formats in the 20th century were: ¾-inch U-matic, Philips VCR, Betamax, VHS and S-VHS, V2000, and the camcorder formats, VHS-C, Video-8 (and Hi8 and Digital8) and DVC. On top of that, there has been an even bigger range of incompatible formats for broadcast and industrial/commercial machines.

As each new generation of videotape format appeared, the formulation of the magnetic tape improved, the complexity of the hardware increased and the precision of writing the video track also increased. Such precision allowed for low cost in tape usage whilst maintaining high performance. This precision challenged what was achievable in technology but the high stakes of market share kept the hardware right at the edge of what was feasible.

Sometimes the manufacturers struggled with the technology. There were occasions, usually in the first pre-production machines, when a tape could only be replayed on the machine on which it was recorded.

The steady improvements in technology with each new format meant increasing sophistication in how the material was recorded and how tape errors were dealt with. That ever-increasing sophistication of format (particularly of the new digital recorders) places a question mark over its long-term survivability. On top of that, the mainstream broadcast video recording formats have been changing more rapidly and in subtly different ways, the more recent they are. We even find today, that simply a different version of a recorder's operating software can render a tape unplayable.

For both the increasing sophistication of the hardware and the decreasing life of each format, archival becomes a mammoth issue amongst the professional broadcasters. So far the problem has not been too bad, but we should be aware that video recorders are designed for today's programmes and for short-term storage to cover a few repeats and not for archival. After all, creating an archive generates cost, not revenue. It is left up to the programme-makers to consider how to preserve their precious material. The problem is that when we talk about archival, we are talking about keeping material for a *very* long time.

Archiving Videotape

The first problem we have is the duration of the videotape format. If a format is superseded, it then becomes essential either to convert it to a new or special archive format, or preserve the tape machine with the videotapes. Preserving the tape machine also implies incurring the expense of maintaining obsolete equipment, keeping a full complement of spare parts and finding knowledgeable service and operational staff.

The second problem comes from the tape itself. Videotapes have only a limited lifetime before they start coming apart. At some stage there just has to be an archiving operation or the material will be lost.

The first such operation has taken place in Britain. The BBC took the decision to undertake a massive archiving operation of most of its Quadruplex tapes onto digital D-3 videotape. The scale of the project was huge, spanning over twenty years of the BBC's entire output. The trigger to start the project was the developing interest in old material for cable, satellite and videocassette, the termination of Quadruplex around the mid-1980s as a transmission format, the inevitable deterioration of the videotapes themselves and – the clincher – the declining number of serviceable machines.

An extreme example of preserving an earlier format is a short fashion programme recorded in the 1950s on the BBC's VERA linear video recorder. The material has been preserved by progressively re-archiving it in the main broadcast videotape formats. It has already been transferred from VERA format to Quadruplex to 1-inch 'C' format to D-3 to Digital Betacam. Watching a VHS copy of the programme, we find it nearly impossible to see which of the more subtle picture faults VERA caused, and which are caused by any of the intermediate generations of formats and copies. As testament to VERA's obsolescence, there are no machines and no known tapes. Only a few individual components of VERA survived as souvenirs and trophies from the team (including one of the empty 20-inch tape reels).

The obsolescence of formats affects us as consumers, though we have been spared for nearly a quarter of a century with JVC's VHS (Video Home System). Initially, VHS rendered its rival, Sony's Betamax, obsolete as a triumph of marketing over technical superiority. Its continued popularity and low cost-of-ownership outweigh its low technical quality against new contenders.

For some of us though, a precious home video recording on Philips VCR *or* V2000 *or* Betamax needs the right machine to play it back. Fundamentally for home video recording, we want something that preserves an entire life history, from 'hatches and matches' to 'despatches'. In the absence of a stable format, this needs to a universally compatible format that can be re-archived without loss of quality.

Dead Formats

The inventors, developers and operators of videotape and videodisc systems throughout the years thought that what they had was, at the time, the 'last word' in such systems. In reality, we can always be sure that something new will come up, rendering the latest and greatest obsolete overnight. It is all too easy to be caught up in the progressive upgrade path and salivate over the next offering from the manufacturers. The epitaph for technology is that it is yesterday's vision, today's market opportunity and tomorrow's junk-pile / museum exhibit / collector's item.

When we stand back from the hardware band-wagon and think of what we *really* want, we see that it differs considerably with what is on offer. Time-shifting a scheduled programme for later viewing using a videocassette recorder is a good example. What we want is only to have some means of being entertained at a time convenient to us, rather than at a time convenient to a programme scheduler. What we do not want is to punch in times or codes into a remote control. We also do not want to think of how much tape is left in the cassette, which tape cassette to use or even what is on the cassette we are about to erase. We may prefer to use either a voice-operated search interface or a computer-like WIMP (windows, icons, a mouse, pull-down menus) interface and select a programme from an on-line TV programme database. This is just one example of the direction we are heading and how far we still have to go.

The End of Media?

When we want to see a movie in our homes, we can watch a scheduled broadcast, or buy a pre-recorded tape or disc and show it in a dedicated playback machine. We do not really need the physical tape or disc. That is after all simply the packaging. We want the movie itself.

This concept has led some visionaries into believing that in the future we will have no need of these playback devices in our homes. As an alternative to conventional broadcasting or 'pay-per-view' programming, we should be able to download all our entertainment from the equivalent of an Internet computer server. 'Pay-per-download' from the Internet is already being pioneered for software and more recently for music. In order to hear the music more than once, we tend to transfer the song onto MiniDisc, CD-R or MP3 for our hi-fi system. However, we have in effect defeated the purpose of data download as we need to store the data locally. The cheapest format is the same on which we would buy the music. Currently, music downloading is a gimmick. It is far better to use the Internet to buy the physical CD, with its glossy descriptive inlay and packaging. Data download on demand is undoubtedly the future, but is dependent on the development of a home mass storage system and high bandwidth low-cost digital communications into the home. Both of these technologies are feasible now and will be affordable soon.

When we order a DVD over the Internet, we can think of this as essentially downloading a movie. It is just that the movie is packaged as between 5 and 10 Gigabytes of data on a plastic disc and the delivery is via mail. With 24-hour delivery, the effective transmission rate from distributor to home is around 500 kilobits per second for a 5 Gigabyte disc. Given the current speed of the Internet via modem, ordering a physical DVD for next day delivery through the Internet is faster (and cheaper) than the download time. Not only that, the DVD is of course a permanent store for the data in the home.

The permanence of DVD and other disc storage technologies belies the temporary nature of the technology to play them back, just as with videotape. The lowest common denominator for the optical discs is the equipment – just as in computing. This is not all that a surprise as the technology for digital video is indistinguishable from computing. As a result, no matter how successful the technology is or how durable the media, it will be obsolete in just a few years.

In technology terms, DVD's successor is already in development. At least one of these optical disc solutions is based on holographic storage. The scientists and engineers are working on a CD-sized disc that stores 125 Gigabytes. One disc would be equivalent to 27 DVDs, around 210 CDs, or even about 15 million Phonovision discs. Technology still moves on, and we must realise that we are paying for the privilege of taking a snapshot of the latest technology rather than an investment for the future. When we look back at all the different domestic videodisc formats, we are looking at an ancestral line on a technological genealogy chart.

When she was five, I showed my daughter her ancestral family tree. Wondering how much she understood of it, I asked her 'What's special about all these people?' She understood more than I thought as she confidently replied, 'Daddy, they're all dead!' In the same way, the videodisc formats that precede the current – Video CD, Laserdisc, RCA's Capacitance Electronic Disc (CED), Matsushita-JVC's Video High Density disc (VHD), TeD's – are also all dead. For videotape, the list of domestic dead formats is much longer, and when the broadcast formats are included, this graveyard for technology looks well populated. When we include the formats that made it to market but never sold and the formats that expended research effort but never made it to market, the list becomes quite incredible.

An Archive for the Future

What will we be watching in 50 years? In 100 years? Or even – and we should consider this very seriously – 1,000 years from now? Just as it is astonishing how some throw-away items, such as comics and cigarette cards, can become highly sought-after collector's pieces a generation later, we simply cannot tell what future generations may want to see. We certainly need to preserve our heritage through news and events, wars and politics, entertainment and sports. Of the rest, we have to make a judgement on what we think might be worth preserving. The limiting factor is the archival material, the archival process and the cost of both. For those items that do not merit special treatment, what survives into the future may simply be what *physically* survives.

Will it be like photography, where the original silver-based black-and-white prints long outlive the more recent dye-based colour prints, or in the motion pictures, where the only movies to survive could be the few, like 'Gone with the Wind' that have received special archival treatment? Will there be such a natural filtering process that selectively erases most of what we know today? The answer to this last point is probably an awesome 'yes'. The reason is information overload, data deluge, the clogging up of the system through the day-to-day additions of new programmes combined with the use of the machines for archival. Inevitably, lack of interest, as perceptions and interests shift with the fashions of the future, will be the root cause of the erasure of today's treasures. It is safe to predict that articles will be written a few hundred years from now that will decry the loss of many classic films through neglect.

Preserving Data

Magnetic recording is not just used for recording video and audio, it is the foundation for computer data storage, and particularly for data archives. This means that equipment obsolescence, format obsolescence and physical degradation affect all computer data stored on magnetic media as much as or even more than it does magnetic tape. In reality, the main danger to computer data is in both the equipment and the format becoming obsolete. Neither normally last long enough for physical degradation to have an effect.

This is especially true of the Phonovision restoration work. When I first started, I used the only affordable storage format for the processed data files: compact audiocassette in FM-encoded digital format. By 1983, I had moved to floppy disc storage using 5¼-inch discs. Each disc was capable of storing 80 kilobytes. Three years later, with a new computer, the data was stored onto 1.4 Mbyte 3½-inch floppy discs. Several years later and the data was stored on 100 Mbyte discs and 648 Mbyte magneto-optical CD-R discs. At each step, the data was migrated onto the new storage formats.

The 3½-inch discs were around for long enough for physical degradation to cause problems. When the 3½-inch discs were being restored from archive, 3 out of around 50 discs, roughly 6% of the discs, had failed after ten years of storage. Fortunately, there were multiple copies and none of the early data had really been lost. The data now resides on multiple copies of CD-R, held at multiple sites. For the original data stored on cassette and on the 5¼-inch floppy discs, neither the format nor the equipment to play them back have stood the test of time.

It is quite ironical that the restored data should be less durable than the original material. The technological archaeological *dig* described in this book may yet be in danger of itself requiring a dig to recover the lost data. The CD-R archive of processed data should last a century or more. However, the equipment to play back the data will almost certainly not be around. Continual data migration onto new formats – known as *archival refresh* – is an answer, but it requires the active participation of someone to undertake the migration. The processes for archival refresh apply just as readily to video material, though the problem is happening now. We already have accrued a vast archive of material on analogue videotape.

A Format for Survival

Which is the best way to archive our video material for posterity? A rolling archival and re-archival process in the digital domain seems to be the answer, but the sheer volume of existing and new material dictates being selective. Guessing what the future will bring, the technology of computing is moving far more rapidly than all the other technologies for television and the answer may lie in the explosion of storage capacity and performance that computing technology brings.

Which of the formats would survive on their own the ravages of time? If an archaeologist a thousand years from now were to discover a museum of television history, what would he find? Of the three types of recordings today – tape, disc and film – there would be no tape material left after ten centuries. The binder, the material that bonds the magnetic dust to the plastic tape, would decay first. After 50 years, the tape would be in a poor state. A few decades later, the tape would be completely unusable. This is the 'expected' life and has yet to be proven. Some claim that tape can last much longer – up to 100 years. Even so, all magnetic tapes and media whether audio, video, analogue or digital will decay leaving the precious recording as a pile of dust.

Film would eventually go the same way. The dyes of the colour film will have started to fade after just a few decades. The film material will eventually decompose like the plastic of the tape itself.

Of the discs, the CD and all its digital derivatives would last well over a hundred years. The problem is most likely to arise from the stability of the bonding material that seals the discs. Even today, through manufacturing faults, we can see what happens when decay sets in.

The polycarbonate material is highly stable but does have a finite life. It may come down to the simplicity in which the material was recorded. If there is no knowledge about the data formats that were used to store the information, it may prove to be very difficult to reconstruct the contents of the discs. It might be that the Laserdisc – essentially a binary (pulse-width modulated) recording of an analogue video signal, complete with timing information, would be the simplest video recording to play back.

First and Last?

The one recording format that should outlast all others is mechanical recording onto shellac and vinyl. Assuming the disc surface remains intact, the information would certainly be the simplest to reconstruct. Even if the disc were broken, careful physical reconstruction of the disc would still allow the information on it to be retrieved.

We can expect the equipment to play back the discs to become obsolete. It would be a major engineering challenge to develop a machine from scratch to handle most of our current digital recording formats. However, our descendants could easily build needle-in-the-groove playback equipment. The simplicity of the gramophone record may make it the most successful format for long duration archives.

Baird's Phonovision discs have lasted since the late 1920s without any noticeable degradation caused by age. If looked after, they will outlast all the broadcast-years of videotape recorded since then. One of the Phonovision recordings has a deep crack that may reduce its life as a complete disc. However, a broken disc can be physically glued back together and the clicks at the joins repaired using computer processing.

The Ultimate Archive

The gramophone recording process has been chosen for the ultimate archive of sounds and images of our home planet. In 1977, two Voyager spacecraft left on a journey to fly-by the outer planets of our solar system and then travel on into interstellar space. Their mission has gone into the history books as the most successful unmanned space exploration.

If extra-terrestrials ever find the probes in the far future, it will be through hope more than chance. With that hope, the spacecraft carried a recording in sound and still pictures of life on Earth, in probably the ultimate time capsule. For recording medium, they chose a mechanically recorded gramophone disc made of gold-plated copper (see Figure 9-16). The best thinking is that

Fig 9-16. The Voyager record (below) and reverse, with pictograms describing how to replay the disc (with supplied cartridge) and what speed to use.
Courtesy of NASA/JPL

this will last indefinitely in the vacuum of space at temperatures close to absolute zero. These discs are already the furthest travelled and will undoubtedly be the best preserved of all our recording media. They are however somewhat inaccessible.

The Voyager engineering team even provided a stylus and cartridge together with pictograms showing how to play back the sound and pictures at the correct speed. If each of the probes or discs is ever found, it will be by our descendants, or whatever we evolve into. They will then have the task of restoring the age-old video, using techniques that may not be all that different to those used for Baird's Phonovision. All we have to do is remember where we pointed the spacecraft for the next few hundred thousand years...

---

[1] 'TV buffs find first BBC TV recording', *The Times*, 25th June 1999, p15. (The reporter omitted the key words 'high definition' between BBC and TV and used 'first' instead of 'earliest-known'.)

[2] BIEHL, M.: Private Communication with the Author, Morehead State University, Kentucky, 25th Jan 2000

[3] BIEHL, M.: *ibid*, 25th Jan 2000

[4] ABRAMSON, A.: 'Zworykin, Pioneer of Television' (Univ of Illinois Press), 1995, Ch.1 p2

[5] MAGOUN, A.: Private communication with the Author, 24th Jan, 2000

[6] ABRAMSON, A.: *ibid*, Ch.7

[7] GRAHAM, M. B. V.: 'RCA and the VideoDisc: the Business of Research' (Cambridge University Press), 1986

[8] MERTZ, P. & GRAY, F.: 'A Theory of Scanning and its Relation to the Characteristics of the Transmitted Signal in Telephotography and Television', *Bell Sys. Tech. Journal*, 13, pp464–515

[9] ABRAMSON, A.: 'The History of Television 1880–1941' (McFarland), 1987, p133

[10] ANON.: *Scientific American*, 17th Nov 1877

[11] BERLINER, E.: 'Gramophone', US Patent 372,786 4th May, 26th Sep, 8th Nov 1887

[12] LANE, B.: '75 Years of Magnetic Recording: 2 – The Dark Years', *Wireless World*, Apr 1975, p162

[13] MULLIN, J. T.: 'Discovering Magnetic Tape', *Broadcast Engineering*, May 1979

[14] SALTER, M.: Private Communication, May 2000

[15] GINSBURG, C.: 'The Development of Ampex Quadruplex', from '25 Years of Video Tape Recording', compiled and edited by D. KIRK, May 1981, for 3M UK Ltd

[16] SAWAZAKI, N., 'Helical Scan: the Early Years', from '25 Years of Video Tape Recording', compiled and edited by D. KIRK, May 1981, for 3M UK Ltd

[17] SALTER, M.: Private Communication, May 2000

[18] HOPKINS, S.: Private Communication, 17th May 1984, regarding Phonovision material for the 1985 mini-series 'Television' (Granada TV)

10 Revising History

'History does not repeat itself

... historians repeat each other.'

Attributed to A. Balfour (1848–1930)

Changing Views

Television has only been around for a matter of decades. You would think then that with such a short history, what there is of it ought to be reasonably correct. However, as we take new views, particularly on who did what, first, we continually re-visit, re-assess and revise the early history of television. These changes are more often to do with developing opinions, driven by either cultural changes or personal and political agendas, rather than new information brought to light. This effect is of course universal and is neither limited to television nor indeed to Britain. The United States of America for example has had its television history re-appraised recently.

The contribution by Philo Taylor Farnsworth to the engineering history of television has been raised from virtual obscurity to US-wide recognition. For decades, the invention and development of the iconoscope by Vladimir Kosma Zworykin had overshadowed Farnsworth's prior invention of an electron-tube camera, called an 'image dissector' (see Figure 10-1). This had been with a measure of good cause as Zworykin's 'iconoscope' was the superior system for broadcast television and became the basis for almost every development of the electron-tube television camera right from its conception in 1925 to the last tube-based camera in the 1980s. The new version of history now recognises that Farnsworth's image dissector was conceived, built and demonstrated before Zworykin's iconoscope.

Technically both approaches were quite different. Both camera solutions converted an image of the scene into an electron image. Farnsworth's solution produced a continuous electron image and swept this over a pinhole aperture. At any one time, this let electrons through from only one part of the image; electrons from the other areas of the image were ignored. The signal passed through a subsequent stage of electron amplification – much like a photo-multiplier.

Zworykin's iconoscope however built up the electrical effect over one complete picture, integrating the electronic charge between scans. A beam of electrons scanned the camera's photosensitive surface, discharging the electron image and producing the television signal. This ability to accumulate the electron image between sweeps of the picture made the iconoscope ideal for broadcast television.

Fig 10-1. The Farnsworth Image Dissector tube.
Courtesy of the Royal Television Society, RTS37-52

Over-zealous Claims

Recent publications praising Farnsworth appear over-enthusiastic in promoting his name.[1] Some of the claims in print make out Farnsworth to be responsible for far more than his primary invention of an electronic television camera. He is claimed to be the man who, '...took all of the moving parts out of televisions and made possible today's TV industry, the TV shots from the moon, and satellite pictures', and 'he is the man who invented that electronic box you are viewing'.[2,3] Along with many other Americans, Neil Postman, the Paulette Goddard Professor of Media Ecology hailed him as being 'the inventor of television'.[4] Farnsworth was without doubt a highly innovative and inventive engineer but neither he, nor any other single person nor any one company deserves such an accolade.

His demonstration of an electronic image in September 1927 to his wife, 'Pem', his brother-in-law, Cliff Gardner, and a business colleague, George Everson, was undoubtedly the first. However, this was not a natural image but a shadowgraph, similar to that already demonstrated by Baird in Britain and Jenkins in the USA in 1924-25. Farnsworth had used a 'hot bright carbon arc lamp' shining directly into the front of his camera tube, with an opaque pattern on a glass slide creating a shadow.[5] Given that this arrangement was not considered true television when used on mechanically scanned systems (as with Baird in 1924), it seems an over-statement of Farnsworth's achievement to call this television.

It took him a further few years of intense development to get to the stage of televising a lit scene, but he was eventually successful. The path had been hard, not just technically, but financially and legally. By the early 1930s, Farnsworth had a working television camera based on his original image dissector patent. Much more important for television history in the USA was that Farnsworth held the key patents on electronic television and in 1934 won a painful legal battle against RCA, the owners of the Zworykin iconoscope, over their validity. Legally at least, Farnsworth goes down in history ahead of Zworykin, despite Zworykin's creation being the success story for broadcast television.

British Television in Perspective

In Britain, there is nothing quite as colourful as the patent battle between Farnsworth and RCA. We have however an interesting parallel to Farnsworth in John Logie Baird. Both started out with a vision of achieving television, both worked on their own to develop their solutions to television and both hit hard times when they came up against a larger corporation. As in the USA, the history of Britain's earliest days of television has also changed over the years. In Britain though, the process has been a steady evolution of opinion on how television developed.

Irrespective of our progressive understanding of a worldwide movement to achieve practical television, John Logie Baird remains the first person to have demonstrated television. Whilst the components for a fully electronic television system were being researched and developed in corporate laboratories, Baird was leading Britain into the new television age.

Far too many have criticised Baird for not developing television into the fully electronic system it became. This is unnecessarily harsh on Baird. From his patents, demonstrations and earlier enterprises, Baird developed and implemented *solutions* to a problem, adapting existing technologies to meet his needs. In Baird, we see an innovative thinker, personally committed to developing television systems for the remainder of his life (see Figure 10-2). If criticism can be directed, the Baird Company did not have the benefit of the research department of EMI at its disposal. Arguably EMI and its Central Research Laboratories held the greatest engineering capability in Britain at the time.

Fig 10-2. Close-up of an engineering spare 12-inch diameter 20-facet mirror drum manufactured by B. J. Lynes Ltd to a design by Paul Reveley for Baird's 1937–38 colour television work. This shows the high engineering quality of Baird's later work. Spinning at 6,000 rpm, the mirrors experienced around 2,000 times the force of gravity. The colour camera and cinema projector worked to a special 120-line two-colour format and was demonstrated only twice, in December 1937 and February 1938.

Courtesy of the Author
From the collection of the NMPFT

The electronic camera took many years to develop. It made its first appearance in the UK in the mid-1930s as an individually handcrafted experimental device. The narrowed view we have today is that EMI succeeded in developing an electronic camera and that the Baird Company failed because they stuck with mechanical television. The story suffers in its abbreviation.

In truth, by the early 1930s, the Baird Company, the newly formed (in 1931) EMI and many other companies were involved in *all* aspects of high definition television. The camera was the key (see Figure 10-3) but equally important was the entire infrastructure for broadcast television – none of which existed at the start of the 1930s. For high definition television, this meant not just the camera, but also the special cables, the signal amplifiers, the vision switching (no 'mixing' in those days!), distribution and, most important of all, a special high frequency transmitter. Without the wideband transmitter, high definition television would remain a curiosity. At the

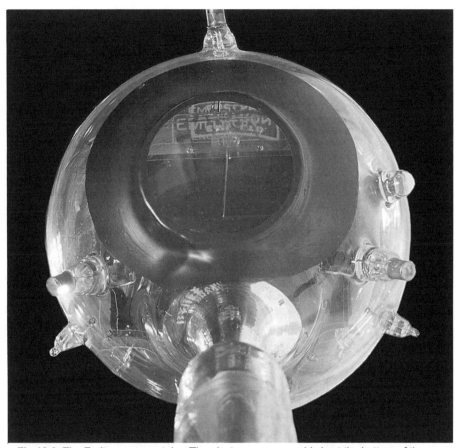

Fig 10-3. The Emitron camera tube. The electron gun assembly is at the bottom of the picture. The dark-rimmed area is an optical window. A lens, not shown, focuses the image onto a photosensitive target just visible as the dark object inside the tube.
Courtesy of the Author
From the collection of the NMPFT

receiving end, there were the self-contained receiving sets for vision and sound. This would be where the manufacturers would *really* make their money.

Those developments in electronic television led to the launch of a high definition television service by the BBC in late 1936. It had exploited the latest developments in electronics to create a service that was able to meet the demands and needs of a broadcast service for at least the next 30 years. In 1936, the scale of the change in cost, scope and systems totally overshadowed all the work that had been done before.

Shifts in Perception

The late 1950s and early 1960s were the halcyon days of television in

Britain. The BBC had extended its broadcast television coverage across most of Britain, the viewing population had grown tremendously since the Coronation in 1953 and BBC Television was establishing itself as the primary media provider over and above the 'Senior Service' – radio. BBC staff, from programme makers to studio managers and engineers, were of the highest quality. They were expected through an established and highly respected heritage to provide a service of excellence.

There was pride in the service and pride in the engineering systems that delivered it. The 405-line monochrome television format had been unchanged since 1936. A quarter of a century of steady development brought with it improved equipment, refining the entire television industry and creating a sense of maturity.

With hindsight, we can now see a low point around this time in the view held of the entire era of 30-line mechanically scanned television. Not even given the grace of just being ignored, television before 405-lines was scorned with vehement attacks on Baird and his works. In 1966, 'The Discovery of Television', a milestone documentary celebrating 30 years of BBC Television, served to pour fuel on the flames.[6] It presented a picture of crudity of equipment (made with 'sealing wax and string'), of amateur performances (the performer 'could barely move a muscle') and an emphasis on the path to electronic television as being the right way and of the mechanical route as being the wrong way. The documentary epitomised the feelings within the television community. Much worse was the influence that these views had on subsequent documentaries.

Why had such a shift in perception occurred? By 1966, the 405-line service had been in service for 30 years, a full generation, with a gap for the War. When these statements were made, the 405-line TV service was mature. Developments were already underway to move completely to 625-line TV and eventually colour. Immersed in such a successful and a durable format, solidly based on valve electronics, it was natural for the documentary makers to take a dim view of the days of television before the start of the 405-line service – a view that persists today amongst many people in the broadcast business.

Triggered in the 1970s by a paper on the history of television, a 'letters' debate opened up in the publications of the Royal Television Society and the Institution of Electrical Engineers regarding certain claims and counter-claims. The debate escalated to form two highly vocal and polarised factions: those who seemed to believe that Baird was the Father of Television, and those who were adamant that he was a charlatan. The correspondence carried the thrust and parry, the serve and return of a contest between professionals vying for recognition of their views. In itself, the battle is a fascinating insight into how people can become so extremely polarised. The sad thing about these two extremes is that neither is now believable. The debate generated more heat from the written exchanges without shedding more light on the subject. Unfortunately, all that has suffered is the image we have of Baird.

The Evolution of Views

It was not always like this. Looking back through publications, we can see the progression in thinking, starting in the 1930s with the advent of alternative systems to that of the Baird Company's. At that time, rapid developments in television technology had created an air of confusion about the future path for television. In 1934, a Committee chaired by Lord Selsdon was tasked with generating recommendations on how to deal with these new developments. Television in Britain was in danger of being driven by the technology – the 'tail' was threatening to 'wag the dog'. The Report from the Committee was extremely favourable about bringing a new service to the public using the results of the latest research and development.

Its recommendations recognised that the then-current low definition BBC Television Service on 30-lines gave entertainment to some, but that the way ahead was towards a high definition public service. The words used in the report were in no way disparaging to the 30-line service and recommended that it continue if it was practical and affordable.

> 'As regards the existing low definition broadcasts, these no doubt possess a certain value to those interested in television as an art, and possibly, but to a very minor extent, to those interested in it only as an entertainment. We feel it would be undesirable to deprive these "pioneer lookers" of their present facilities until at least a proportion of them have the opportunity of receiving a high definition service. On the other hand, the maintenance of these low definition broadcasts involves not only some expense, but also possibly considerable practical difficulties. We can only, therefore, recommend that the existing low definition broadcasts be maintained, if practicable, for the present...'

Despite this, the 30-line service was terminated in 1935 – almost a year before the high definition service started up. The restored amateur video recordings of the 30-line service show us that the programmes were designed around the limitations of the low definition system. Operating a 30-line service alongside the high definition service would realistically have meant two sets of studios and two sets of production costs. Inevitably, the low definition system was dropped.

Even after the Second World War, the low definition service received favourable comments and even praise. 'Considering their low definition – 30 lines compared with the 405 lines put out from Alexandra Palace today – Baird's pictures were surprisingly clear, in spite of the flicker…'.[7]

In 1947, Roy Norris used the obsolete mechanical 30-line system to describe and illustrate how modern television worked. He said that the Nipkow disc Televisor '…employed principles which are still fundamental'.[8] This is a far cry from the later damning indictments of the 1960s.

John Swift of the BBC wrote a history of television in 1950 called 'Adventure in Vision'. It was subtitled 'the first twenty-five years of television' recognising that television started with John Baird's experiments. Swift's book gave credit to the contribution that Baird and the other pioneers had made. Eleven years later, Gordon Ross wrote another history of television entitled 'Television Jubilee'. The subtitle 'The Story of 25 years of BBC Television' was an indication of a shift in perception to a post-1936 world of television.

By 1976, the 40th anniversary of the BBC's 405-line television service, the 30-line television period achieved a revival of support. Bruce Norman's documentary, 'The Birth of Television', and accompanying book, 'Here's Looking at You', gave a fair account of the 30-line days in context with the later 405-line service, though based heavily on personal reflection by those who made it. Some of the image persisted. The impression still given was of amateur productions on a crude experimental system, with due emphasis on the history-making aspects of Baird's work and the BBC's involvement in 30-line television. One of the primary supporters of that shift in emphasis was Tony Bridgewater.

'Just a Few Lines'

Tony Bridgewater, Chief Engineer of BBC Television from 1962-68, was a late supporter of the mechanical TV era. He had a background extending throughout almost all of Britain's television engineering history and was one of the three engineers responsible for the BBC's 30-line television service. As a result, he was probably the best person to give an accurate appraisal of the 30-line period.

After his retirement, Bridgewater actively attempted to create a more balanced view of the 30-line period and television history in general. He was instrumental in bringing to light the contribution by A. A. Campbell Swinton.[9] He campaigned against his old employers, the BBC, to drop the word 'experimental' from the plaque commemorating the BBC's 30-line service in Broadcasting House. His final publication was a monograph on the 30-line period 'Just a Few Lines'.[10] Made over half-a-century after the service was closed down, the monograph captures the 30-line period as no previous publication had done. Those who were directly involved praised the monograph for its accuracy. The rest of us who had been brought up on the accepted view read the monograph and thought that time may have added a certain rosy tint to Bridgewater's spectacles (see Figure 10-4).

Fig 10-4. Connie King on 30-lines (left) and in real-life (right). The quality of the image in the television studio was invariably superior to the image received in the home. This was largely due to using the medium waveband for video transmission. Traditionally, criticism on video quality has been levelled at the 30-line format. However, the weak link in the transmission chain was the existing broadcast radio network.

From originals courtesy of R. M. Herbert

The restored amateur recordings of BBC transmissions have provided the evidence to support Bridgewater's view. Whereas the reviews, books and magazine articles written in the 1930s could be regarded as being influenced by the novelty of the new system, there is no denying the evidence of real video recordings made at the time. For as long as I knew Bridgewater, he had wished that there had been recordings made of the BBC 30-line transmissions. Just before he died, I had the greatest pleasure in making him one of the first people to see the restored Paramount Astoria Girls performance from 1933. Unable to speak through his illness, though still very alert, he jotted down 'Very nostalgic'. Bridgewater died a few months later. The BBC and the television industry had lost one of their most highly respected engineers and one whose career spanned the world's first regular scheduled television service in 1929, right through to the introduction of colour television throughout Britain from 1967.

Bridgewater's monograph received only specialist circulation. As a result, it had little impact on the public's view. The BBC's first Television Service remained 'experimental', the productions were still stilted and amateurish and Baird was still 'the inventor of television' through using a 'Pile of Junk'. The image presented by the 'Discovery of Television' documentary in 1966 prevailed. At one point, the commentator, Derek Hart, says '...the modern television system contains nothing, not one single piece of equipment or idea originated by John Logie Baird'. This is an unnecessary and unjust knife in the back for Baird. To appreciate the comment we need to understand that it was said genuinely in the light of that sense of engineering maturity that prevailed in the 1960s (see Figure 10-5). However, with the rapid progress in digital electronics from the 1970s onwards, the period of perceived maturity gave way to a time of rapid development and enhancements of the systems for television that has yet to show signs of slowing down.

The documentary's comment related Baird's system of 1926–36 with the 'modern' systems of over 30 years later, in 1966. Some 30-odd years beyond that, in early 2000, you could go out to a high street shop and buy a device that has almost no technology from the 1960s in it. Today's Digital Video Camcorder is as remote from the 1960s technology as 1960s television was from that of the 1920s and 1930s. Looking back on that documentary, the harshest view we should have is that the maker suffered from not just a lack of, but also a failure of imagination.

A Digital Viewpoint on Television

To date, the views of television history have been based on research and writings made in a time of stability for the technology. Times have changed and television and its associated technologies are in a stage of rapid development. The changes are so rapid that it can be difficult to get an overall view of where we are and where the technology is leading us.

In late 1998, the BBC launched a Digital Television service, starting a new era in television's history. That is what history will probably attempt to record in sound-bite fashion. Alternatively, history could record that after 25 years of migrating their studio hardware – their cameras, recorders and editing equipment – from analogue to digital, the major television networks around the world improved the communications hardware for broadcasting and reception of television with digital technology. In Britain in 1998, digital television made only that small step.

Broadcasting in Britain needs to embrace further improvements to the service. Just having the additional quality of digital broadcasting is insufficient for the public to go and buy the latest hardware en masse – particularly if they are savvy enough to understand that further changes are due – even overdue.

Already the extra features on DVD movies hint at the next steps: Internet-like mark-up languages and Internet browser solutions to replace ageing Teletext systems, six-channel Dolby Digital and DTS audio to replace existing NICAM digital stereo, interaction with the programme material and multiple camera angles. In addition, we are transitioning to digital video recorders for capturing and time-shifting all of this at broadcast quality. These changes are nowhere near what we have already seen in television in Britain: NICAM digital stereo audio in 1986, Teletext in 1974, PAL colour in 1967, 625-line in 1964 and, above all, the high definition (405-line) television service opened by the BBC in 1936.

Fig 10-5. An RCA CTC11 NTSC television receiver circa 1960.

Courtesy of the Author

End of the Tube?

Yet, even at the beginning of the 21st century, at the heart of our television set receiving digital images from satellite or DVD, is a relic of a bygone age – a thermionic valve. This valve – the cathode-ray tube, CRT, or just the *Tube* – houses glass and metal in a vacuum with high voltage electron beams that continually paint our television picture in a technology solution from one hundred years ago, at the start of the 20th century.

This single remaining valve in our television receivers lingers on chiefly due to it remaining more cost-effective than any alternative. There is of course no nostalgia associated with keeping the tube in our receivers. It is complex to manufacture, requires high voltages and analogue circuitry.

As soon as a cheaper and better alternative becomes available, we will see the last of valve-technology. We would then have a television system comprising chip-based cameras, digital communications sending encoded blocks of data rather than lines, all received and shown on flat-screen displays. Not too far in the future we will have a television system that will have far more in common with the thought-experiments of people in the late Victorian period than that of any of the early 20th century pioneers.

We have reached the unique vantage point in time of being able to look *back* at analogue television and see it as history. We see it now as yesterday's technology. Those engineering developments that gave rise to the arrogance of the 1960s appear now to be no more than stepping stones in the development of television, a development that will continue with that of the supporting technologies. This vantage point allows us to gain a far better appreciation of the impressive developments made by the early television pioneers. Of these, John Logie Baird (see Figure 10-6) was the most prolific, the most innovative and in Britain certainly, the most exceptional.

Baird has easily earned the acclaim of being Britain's foremost television pioneer. If he were only known for giving the world's first demonstration of television, then that would be sufficient to secure his place in history. Uniquely amongst the television pioneers, Baird developed, demonstrated and patented almost every aspect of television, including colour, infrared and stereoscopy. He introduced and funded his own television service, even paying the BBC for the use of their transmitters. His 30-line system was adopted, and hence sanctioned, by the BBC for their first television service.

The Legacy

The Phonovision recordings give us a deep insight into not only Baird's experiments but also most of the problems he encountered along the way. We understand the errors in making his experimental Nipkow discs, the problems in trying to get his amplifiers stable, the power of his electric motors, and much more. The professional quality of the Phonovision disc recordings has allowed the 'signature' of the faults in the laboratory to be extracted and interpreted.

The lessons learned from the domestic recordings however are quite different. The faults in the Baird and BBC studio equipment are masked by the poor quality of the domestic equipment. They do however reveal a great deal about the quality of the performance, the production, the lighting and the camera-work. This is the first time we have been able to do this since the transmissions were made. Additionally, we can compare the techniques

Fig 10-6. John Logie Baird – Britain's foremost television
pioneer
Processed from 'Television', Dinsdale, 1928

with those of today, study what production techniques were used back at
the very beginnings of broadcast television and see how they fit into the
development of the medium.

It may seem rather strong to say that the results of these recordings have
re-written history. The history books that we have depended on have
attributed a blanket low quality of service to the entire 30-line period. They
also describe the technical quality of broadcast 30-line TV as generally
crude, and more in relation to the far earlier experimental efforts of John
Logie Baird. As such, the domestic recordings now provide us with a
reference to which those historians never had access. This is why their
writings need to be re-calibrated.

Conclusion

From the movement of a needle in a groove, and the use of computer-based
signal and image processing, we now have a wealth of knowledge and
video material from the very beginnings of television in Britain. From

Phonovision, which Baird himself regarded as his least successful experiment, we have details of the vision format, the equipment he used, the problems he encountered, and an understanding as to why Phonovision was a failure. However, the computer can correct for these problems and we can now get some idea of what Baird's studio was like and what he was trying to achieve. The later private recordings of BBC 30-line Television tell us far more about the pioneering production techniques, the camerawork and the lighting than we could ever get from the written word.

That we have video recordings made during the historic development of Baird's 30-line television system and during the first BBC Television Service is a wonder in itself. That we can now see these pictures as they would have appeared then, is of considerable importance to television's history. That we now know what equipment Baird used and understand the problems he encountered during his development of Phonovision goes beyond what the history books have told us. That we can now properly appreciate the high production quality of BBC 30-line Television in the early 1930s goes some way to correcting long-established dogma. Moreover, that we can tell all of this from studying the movement of a needle in a groove – that alone makes the story worth telling.

---

[1] SCHATZKIN, P.: 'The Farnsworth Chronicles', Web-site 1999

[2] 'National Inventors Hall of Fame', Web-site 1999

[3] WPTA Fort Wayne, Web-site 1999

[4] POSTMAN, N.: Time magazine's 100 most important people of the 20th century, 1999

[5] SCHATZKIN, P.: 'The Farnsworth Chronicles', Web-site 1999

[6] 'The Discovery of Television', Mullard Ltd in association with the BBC, transmitted 3rd Nov 1966, producer A. MILNE.

[7] CHEVIOT, G.: 'The story of television', The World Radio and Television Annual, 1947

[8] NORRIS, R. C.: 'Television Today' (Rockcliff), 1947, p14

[9] BRIDGEWATER, T. H.: 'A. A. Campbell Swinton, FRS' (Royal Television Society monograph series), 1982

[10] BRIDGEWATER, T. H.: 'Just a Few Lines' (British Vintage Wireless Society)

Annex

Derivation of Aspect Ratio from Arc-scanning

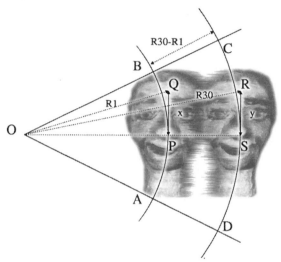

In the figure above, ABCD represents the scanned area on the Nipkow disc with angle AOB being $2\pi/30$ – assuming a single spiral of 30 apertures per revolution of the disc. Arcs AB and CD are the paths followed by lines 1 and 30 respectively. For small angles, the height of the object at these extremes are:

$x/R_1 = \sin f$

$y/R_{30} = \sin g$

where f is angle POQ, or $F_1*2\pi/30$, g is angle SOR, or $F_{30}*2\pi/30$, and F_1 and F_{30} are the fractions of the length of line for the respective lines directly measured from the best fit graph below.

$F_n = 0.36154 - 0.00117*n$, where n is the line number 1 to 30

such that $F_1 = 0.36037$ and $F_{30} = 0.32644$

As the dummy head maintains constant height across the frame, then $x=y$. We can then gather the equations to derive an expression for the width, W, of the image.

$W = R_{30} - R_1$

$= R_1*(a-1)$

$= -R_{30}*(1/a-1)$

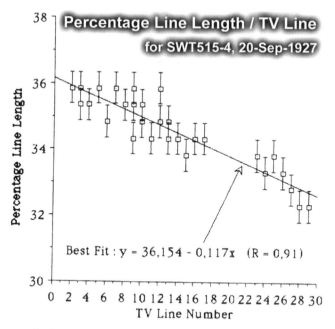

where $a = \sin f / \sin g$

Now the height, H, of the scanned area can be expressed by the arc-length and differs on lines 1 and 30.

$H_1 = R_1 * 2\pi/30$

$H_{30} = R_{30} * 2\pi/30$

The aspect ratio, AR, is defined as the ratio of height to width of the scanned area, or the average of the instantaneous values for the aspect ratio on lines 1 and 30.

$AR = -\pi * ((1+a) / (1-a)) / 30$

Using the values of F_1 and F_{30} calculated from the regression analysis we get a value for AR, the aspect ratio, of **2.12:1** (vertical : horizontal) in the middle of the frame. This is within 10% of the published value of 2.33:1 (or 7:3).

Bibliography

Author's Comment

The books listed here are by no means an exhaustive list, but are ones which have contributed value to the author's development of the work in this book.

The quality of retrospective books on early television is variable as they chiefly incorporate primary material wrapped up in contemporary commentary. For historical accuracy and research, these books should be used as channels to access the primary material. The best sources of primary material are the patents, the engineering books (listed below, especially Shiers, 1997) and the engineering periodicals: 'Television' (from March 1928), 'Practical Television' (from September 1934) and 'Wireless World'.

Recent Publications on Early Television

ABRAMSON, A.: 'The History of Television, 1880 to 1941' (McFarland & Co. USA), 1987, ISBN 0899502849

ABRAMSON, A.: 'Zworykin, Pioneer of Television' (University of Illinois Press USA), 1995, ISBN 0252021045

BARNOUW, E.: 'Tube of Plenty: the evolution of American Television' (Oxford University Press), 1975, 1982, 1990

BRIDGEWATER, T. H.: 'Just a Few Lines' (British Vintage Wireless Society), 1992

BURNS, R. W.: 'British Television: The formative years' (IEE History of Technology Series Vol. 7, Peter Peregrinus Ltd), 1986, ISBN 0863410790

BURNS, R. W.: 'Television: an international history of the formative years' (IEE History of Technology Series Vol. 22, IEE), 1998, ISBN 0852969147

EVERSON, G.: 'The Story of Television: the life of Philo T Farnsworth' (Norton 1949, and Arno Press 1974), ISBN 0405060424

HERBERT, R. M.: 'Seeing by Wireless' (P.W. Publishing), 2nd edition, 1997, ISBN 1 874 110 15 8

NORMAN, B.: 'Here's Looking at You: the Story of British Television 1908–39' (BBC and the Royal Television Society), 1984, ISBN 0563201029

SHIERS, G.: 'Early Television: A Bibliographic Guide to 1940' (Garland Publishing USA), 1997, ISBN 0824077822 (this publication contains references only)

Signal Processing

GODSILL, S. J. & RAYNER, P. J. W.: 'Digital Audio Restoration' (Springer-Verlag), 1998, ISBN 3540762221

RABINER, L. R. & GOLD, B.: 'Theory and Application of Digital Signal Processing' (Prentice-Hall), 1975, ISBN 0139141014

OPPENHEIM, A. V., SCHAFER, R. W. & BUCJ, J. R.: 'Discrete-Time Signal Processing', 1999, ISBN 0130834432

Image Processing

ANDREWS, H. C. & HUNT, B. R.: 'Digital Image Restoration' (Prentice-Hall), 1977, ISBN 0132142139

GONZALEZ, R. C. & WOODS, R. E.: 'Digital Image Processing (World Student Series)' (Addison-Wesley World Student Series), 1994, ISBN 0201600781

HUANG, T. S.: 'Advances in Computer Vision and Image Processing Vol 2: Image Enhancement and Restoration' (JAI Press), 1986, ISBN 0892324600

HUANG, T. S.: 'Advances in Computer Vision and Image Processing Vol 3: Time-varying Imagery Analysis' (JAI Press), ISBN 0892326352

KASTSAGGELOS, A. K.: 'Digital Image Restoration' (Springer-Verlag), ISBN 3540532927

PRATT, W. K.: 'Digital Image Processing' (John Wiley & Sons), 1991, ISBN 0471857661

Television and Video Recording Systems

BROWN, B.: 'Amateur Talking Pictures and Recording' (Pitman), 1933

DANIEL, E. D., MEE, C. D. & CLARK, M. H. (eds): 'Magnetic Recording: the First 100 Years' (IEEE), 1998, ISBN 0780347099

KIRK, D.: '25 Years of Videotape Recording', for 3M UK Ltd, 1981

LUTHER, A. C.: 'Video Recording Technology' (Artech House), 1999, ISBN 0890062757

MATTHEWSON, D. K.: 'Revolutionary Technology' (Butterworth), 1983

PEARSON, D. E.: 'Transmission and Display of Pictorial Information' (Pentech Press), 1975, ISBN 0727321013

ROBINSON, J. F.: 'Videotape Recording. Theory and Practice' (Focal Press), 1975

Historical Television Books

ABRAMSON, A.: 'Electronic Motion Pictures' (University of California Press 1955, reprinted Arno Press 1974), (this is the first 'history' book to recognise that J. L. Baird made the first television recordings)

BARNOUW, E.: 'A Tower in Babel (History of Broadcasting in the US)' (Oxford University Press), 1967, (along with 'The Golden Web' and 'The Image Empire' forms what was considered to be the definitive history of US broadcasting)

BARTON CHAPPLE, H. J.: 'Popular Television: Up-to-date principles and practice explained in simple language' (Pitman), 1st edn, 1935

CAMERON, J. R.: 'Radio and Television' (Cameron), 1935

CAMM, F. J.: 'Television and Short-wave Handbook' (Newnes), 1st edn, 1934

DINSDALE, A.: 'Television' (Pitman), 1926 (1st edn is first ever book devoted to television), 1928 (2nd edn greatly expanded)

DINSDALE, A.: 'First Principles of Television' (Chapman & Hall), 1932

DOWDING, G. V. (ed): 'Book of Practical Television' (Amalgamated Press), 1935

FELIX, E. H.: 'Television, Its Method and Uses' (McGraw-Hill Inc USA), 1931

HUTCHINSON, R. W.: 'Television up-to-date' (University Tutorial Press), 1935 (1st edn), 1937 (2nd edn)

JENKINS, C. F.: 'Radio Pictures' (Jenkins USA), 1925

LARNER, E. T.: 'Practical Television' (Ernest Benn), 1928

MOSELEY, S. A. & BARTON CHAPPLE H. J.: 'Television, Today & Tomorrow' (Pitman), 1931 (2nd edn)

NORRIS, R. C.: 'Television Today' (Rockliff), 1947 (describes Baird's war-time colour TV work)

REYNER, J. H.: 'Television Theory and Practice' (Chapman & Hall), 1934

ROSS, G.: 'Television Jubilee: The Story of 25 years of BBC Television' (W H Allen), 1961

SHELDON, H. H. & GRISEWOOD, E. N.: 'Television: Present methods of picture transmission' (Van Nostrand USA), 1929

SHIERS, G. (ed): 'Technical Development of Television' (Arno Press), 1977, (Collected Papers)

SWIFT, J.: 'Adventure in Vision: the first 25 Years of Television' (John Lehmann), 1950

TILTMAN, R. F.: 'Television for the Home' (Hutchinson & Co.)

WEST, Capt. A. G. D. (ed): 'Television Today' in 2 volumes (Newnes), 1935

Biographies of John Logie Baird

BAIRD, J. L.: 'Sermons, Soap and Television' (Royal Television Society), 1988 (originally titled by Baird as 'Sermons, Socks and Television', this is Baird's autobiography as dictated in 1941)

BAIRD, M.: 'Television Baird' (HAUM), 1973, ISBN 0798600527 (An account of Baird's life by his wife using extracts from his then-unpublished autobiography)

BURNS, R. W.: 'The Life and Times of John Logie Baird' (IEE), to be published 2000

MOSELEY, S.: 'John Baird: the Romance & Tragedy of the Pioneer of Television' (Odhams Press), 1952 (A romanticised account of Baird's life by his ex-business partner and manager using extracts from his then-unpublished autobiography)

TILTMAN, R. F.: 'Baird of Television: the Life Story of John Logie Baird' (Seeley Service & Co), 1933 (A contemporary romanticised account of Baird's life to 1933)

Author's Publications

MCLEAN, D. F.: 'Television's First Gags by Stookie Bill', *New Scientist*, 20th Oct 1983

MCLEAN, D. F.: 'Using a Micro to process 30-line Baird television recordings', *Wireless World*, Oct 1983, pp66–70

MCLEAN, D. F.: 'Moving Pictures from Wax', *Radio and Electronics World*, Feb 1984, pp61–63

MCLEAN, D. F.: 'Computer-based analysis and restoration of Baird 30-line television recordings', *Journal of the Royal Television Society*, *22/2*, Apr 1985, pp87–94

MCLEAN, D. F.: 'The Recovery of Phonovision', Third International Conference on Image Processing and its Applications, University of Warwick, 18–20th July 1989, *IEE Conf Pub 307*, pp300–304

MCLEAN, D. F.: 'Sounds of the 20's', *Electronics Weekly*, No 1875, 2nd Sep 1998, pp26–27

MCLEAN, D. F.: 'Dawn of Television', *Electronics World*, **104**, No 1749, Sep 1998, pp745–749

MCLEAN, D. F.: 'Restoring Baird's Image', *Electronics World*, **104**, No 1750, Oct 1998, pp823–829

MCLEAN, D. F.: 'First Frames', *Electronics World*, **104**, No 1751, Nov 1998, pp943–946

MCLEAN, D. F.: 'Looking In...', *Electronics World*, **104**, No 1752, Dec 1998, pp1031–1034

MCLEAN, D. F.: 'Digital Visions', *Electronics World*, **105**, No 1754, Feb 1999, pp160–165

Index